The Engineering Capstone Course

Harvey F. Hoffman

The Engineering Capstone Course

Fundamentals for Students and Instructors

Harvey F. Hoffman
Fairfield University
Fairfield, CT, USA

Additional material for this book can be downloaded from http://extras.springer.com

ISBN 978-3-319-05896-2   ISBN 978-3-319-05897-9 (eBook)
DOI 10.1007/978-3-319-05897-9
Springer Cham Heidelberg New York Dordrecht London

Library of Congress Control Number: 2014942312

© Springer International Publishing Switzerland 2014
This work is subject to copyright. All rights are reserved by the Publisher, whether the whole or part of the material is concerned, specifically the rights of translation, reprinting, reuse of illustrations, recitation, broadcasting, reproduction on microfilms or in any other physical way, and transmission or information storage and retrieval, electronic adaptation, computer software, or by similar or dissimilar methodology now known or hereafter developed. Exempted from this legal reservation are brief excerpts in connection with reviews or scholarly analysis or material supplied specifically for the purpose of being entered and executed on a computer system, for exclusive use by the purchaser of the work. Duplication of this publication or parts thereof is permitted only under the provisions of the Copyright Law of the Publisher's location, in its current version, and permission for use must always be obtained from Springer. Permissions for use may be obtained through RightsLink at the Copyright Clearance Center. Violations are liable to prosecution under the respective Copyright Law.
The use of general descriptive names, registered names, trademarks, service marks, etc. in this publication does not imply, even in the absence of a specific statement, that such names are exempt from the relevant protective laws and regulations and therefore free for general use.
While the advice and information in this book are believed to be true and accurate at the date of publication, neither the authors nor the editors nor the publisher can accept any legal responsibility for any errors or omissions that may be made. The publisher makes no warranty, express or implied, with respect to the material contained herein.

Printed on acid-free paper

Springer is part of Springer Science+Business Media (www.springer.com)

I dedicate this book to my grandchildren Sydney, Ella, Ruby, Aviva, and Idan. I hope that they lead inspired, rich, and satisfying lives. I wish that all the projects in which they involve themselves in their lifetime are successfully completed and meet their expectations.

Preface

> *Life can only be understood backwards; but it must be lived forwards.*
> —Soren Kierkegaard
>
> *Whether you think you can or think you can't, you're right.*
> —Henry Ford

Undergraduate and graduate students enrolled in an engineering program are exposed to a great many engineering tools, methods, and theory but engineering practice is not a theoretical profession. The realities of the engineering world dictate that technological skill is only a part of what an engineer needs to succeed. Engineering practice in industry consists of more than the practical application of science to solve problems. Knowledge of mathematics and engineering theory is necessary but not sufficient to be a successful engineer.

The seminal engineering education Green Report [1] stated

> In today's world and in the future, engineering education programs must not only teach the fundamentals of engineering theory, experimentation and practice, but be RELEVANT, ATTRACTIVE and CONNECTED:
>
> RELEVANT to the lives and careers of students, preparing them for a broad range of careers, as well as for lifelong learning involving both formal programs and hands-on experience;
>
> ATTRACTIVE so that the excitement and intellectual content of engineering will attract highly talented students with a wider variety of backgrounds and career interests particularly women, underrepresented minorities and the disabled and will empower them to succeed; and
>
> CONNECTED to the needs and issues of the broader community through integrated activities with other parts of the educational system, industry and government.

The Green Report encouraged engineering educational institutions to incorporate a broad framework for engineering curricula that includes the following:

Team skills, including collaborative, active learning;
Communication skills;
Leadership;
A systems perspective;

An understanding and appreciation of the diversity of students, faculty, and staff;
An appreciation of different cultures and business practices, and the understanding that the practice of engineering is now global;
Integration of knowledge throughout the curriculum;
A multidisciplinary perspective;
A commitment to quality, timeliness, and continuous improvement;
Undergraduate research and engineering work experience;
Understanding of the societal, economic, and environmental impacts of engineering decisions; and
Ethics.

Many engineering colleges and universities introduced the senior design or capstone course to help ensure that they met the challenges set forth in the Green Report.

My motivation for writing this book is to introduce students to an engineering industry model that supports the practices demanded of new engineering school graduates by business and industry. I believe I have a useful and meaningful perspective having worked in industry for 30 years before becoming a full-time professor. During this time I held positions ranging from research and development engineer, group manager, project manager to general manager. In the aerospace field I contributed to activities on the space shuttle and the Titan rocket. I participated in the development and production of radar and artillery weapon systems; and analytical instruments, systems, and services for radiation detection and radiation monitoring. I agree with the comments in the Green Report, which suggest that successful engineering professionals in industry rely on a combination of technical depth, business fundamentals, communication competencies, an appreciation for societal issues, and interpersonal skills.

Some new engineering college faculty members do not have business and industry work experience. Like all teachers their goal is to help students understand engineering concepts and help them to apply, analyze, and synthesize, create new knowledge, and solve problems. However, effective engineering teaching also requires engineering teachers to prepare students for the realities and demands of business and industry—whether or not they have familiarity with the issues in this domain. This book will guide instructors and challenge students to consider not only technical engineering considerations but also to take into account practical business concepts that influence a project and without which a project will never get off the ground. The capstone course forces students to be creative and learn design not solely from theory but from practical applications. The capstone hands-on practical experience helps prepare students for the transition to the workforce and can reinforce their value.

Boeing Aircraft took the lead in highlighting the need for developing competent engineers in their university supply chain of talent. It published a list of desired engineering attributes with the hope of helping engineering education programs to better align themselves with strategic employer needs [2, 3].

Desired Attributes of an Engineer (from Boeing Aircraft)

- A good understanding of engineering science fundamentals
  - Mathematics (including statistics)
  - Physical and life sciences
  - Information technology (far more than "computer literacy")
- A good understanding of design and manufacturing processes (i.e. understands engineering)
- A multidisciplinary, systems perspective
- A basic understanding of the context in which engineering is practiced
  - Economics (including business practice)
  - History
  - The environment
  - Customer and societal needs
- Good communication skills
  - Written
  - Oral
  - Graphic
  - Listening
- High ethical standards
- An ability to think both critically and creatively—independently and cooperatively
- Flexibility. The ability and self-confidence to adapt to rapid or major change
- Curiosity and a desire to learn for life
- A profound understanding of the importance of teamwork.

This list contains the important competencies that Boeing believes that students need to operate in a modern engineering environment. Boeing also observed [3] that there was as much ignorance of the realities of academe on the part of industry as there is on the part of faculty about industry. This book represents an effort to assuage this problem by preparing students for the step into industry using the capstone course as the pathway.

The undergraduate or graduate capstone courses provide students with an opportunity to apply concepts and tools studied in the engineering program to the solution of a "real-world" problem. Often undergraduate institutions refer to the course as a senior design project. This book uses the terms senior design or capstone course interchangeably. Both courses cover similar material. Students work in small groups on a problem proposed by their team or a faculty mentor. They receive guidance and mentorship from faculty who have experience in completing similar activities in industry or academe. The capstone course requires students to integrate knowledge gained from previous courses including project management, business, and engineering disciplines, as well as the practical experience that they may have gained from industry. The objective is not to develop complex

engineering products or services, but rather to introduce and understand the multidimensional processes by which a design is selected, guided, and controlled. The instructor creates an active learning environment. The capstone course implements John Dewey's (one of the most significant educational thinkers of the twentieth century) concept of education by having a curriculum that is relevant to students' lives. Dewey saw learning by doing and the development of practical skills as crucial to education. This start-to-finish project course involves students completing the following "learning by doing" steps:
- Identifying a need for a product, service, process, or system
- Creating a team and working effectively together towards common goals
- Generating a project proposal
- Preparing a design
- Developing the product, service, process, or system
- Testing the product, service, process, or system
- Examining the business viability aspects of the project
- Preparing and delivering reports and presentations.

Capstone projects require students to get up-to-speed quickly on a variety of content areas; enhance key skills such as project management and teamwork; develop competency in gathering, analyzing, and reporting data; understand the relationship between engineering and the business environment; communicate with stakeholders; and construct a product or service. It represents an opportunity for students to interweave their learning in all these areas and to do so in real time, in an unpredictable, complex, and real-world environment.

After many years in industry as an engineer and engineering manager, I returned to the academic world as both a teacher and college administrator. For the past 5 years, I taught a graduate engineering capstone class at Fairfield University wherein I tried to share with my students' ideas that I gained dealing with industry's expectations about the engineering design and development process. The course requires that students integrate important competencies and perspectives to which they have been exposed. The course attempts to facilitate the student's transition from the academic to the professional world by drawing upon all of their skills and abilities. The course challenges the student in technical design; project management; teambuilding; oral and written presentations; time management; preparation and analysis of a schedule; budget preparations; resource management; business decision making; and the consideration of societal impacts on their work. Students work toward achieving team consensus on a diverse set of technical and business data compiled by team members. They synthesize the information in written and oral reports, while resolving conflicts that may arise. This book contains material that serves as the basis for the 1-year course that I mentor.

I evaluate students primarily on their performance in articulating the design, business, and management aspects of the project rather than on the performance of the final product or service. The technical and management processes learned in this class are most important. It is likely that the completed product or service will meet a basic requirement, but the most important course outcome is an understanding of the processes used in industry to complete a project. Learning and understanding these processes is the focus of the capstone course and this book. The typical capstone course follows the five phases shown in Table 1.

Preface   xi

Table 1 Five phases in the capstone course

Course phase	Technical requirement	Skill development	ABET outcome
Team selection	Understand the concept: Delight the customer	Investigation: Determine a need for a product, service, process, or system Identify potential team members with an interest in working on a topic associated with the identified "need" Understand and apply team member's strengths to the problem solution	d, f, g
Project selection	Research alternatives Confirm project need Relate client/customer's needs to a project specification Understand the project's societal impact Select the best alternative Investigate competitive products and services including intellectual property issues (patents, licenses, copyrights, and trademarks)	Organizing and preparing presentations Researching alternatives Oral presentation skills Resolving conflicts that arise in team settings Intellectual property research	g, h, j
Proposal preparation and presentations	Organize an approach to a design Identify objectives, constraints, and criteria Identify data needs and collection methods Evaluate alternative approaches Investigate competitive products and services Resolve intellectual property research "loose ends" Summarize the overall approach with a functional block diagram Relate client/customer's needs to the project specification Prepare a specification Understand the societal, environmental, political, ethical, health, and safety aspects of the project. Agree on final project features and agree on what not to do	Presentation skills—ranging from speaking before an audience to the use of audio-visual aids and application software Project management skills—including collaborative work and communication within and outside the team Writing technical memos and reports Team management skills—achieving team consensus on diverse technical input compiled by different team members and synthesizing them in written and oral reports Resolving conflicts that arise in team settings Meetings with prospective clients, users, and vendors	a, e, f, g, h, j

(continued)

Table 1 (continued)

Course phase	Technical requirement	Skill development	ABET outcome
Project development	Learn new technical skills "on-the-fly" to support the project Learn to make appropriate technical, cost, and feature tradeoffs Make schedule adjustments Explain schedule variance Specification adjustments Prepare a bill of materials Develop project test methods to confirm the product/service meets the specification Succinctly discuss both technical and nontechnical engineering project issues Product, service, process, or system testing by using a test verification matrix Understand the human and regulatory side of the design process Understand and use appropriate industry standards	Presentation skills, ranging from speaking before an audience to use of audiovisual aids and application software Project management skills—including collaborative work Learning to defend team's work before outside evaluators Team management skills Learn to recognize and mitigate risks Quickly learn about unfamiliar topics Develop the ability to discuss both technical and nontechnical engineering project issues with technical and nontechnical people	a, b, c, e, f, g, i, k
Final report, presentation, and demonstration	Summarize project highlights Analyze and explain reasons for technical differences between the target specification and the final product or service Analyze and explain schedule variance Analyze and explain budget variance Discuss actual outcomes compared with performance expectations and budget.	Presentation skills, ranging from speaking before an audience to use of audiovisual aids and application software Learn to defend the team's work in front of stakeholders and outside evaluators	e, f, g, k

The Accreditation Board for Engineering and Technology (ABET) [4], outcomes referred to in Table 1 as listed in Criterion 3 are the following:
(a) an ability to apply knowledge of mathematics, science, and engineering
(b) an ability to design and conduct experiments, as well as to analyze and interpret data
(c) an ability to design a system, component, or process to meet desired needs within realistic constraints such as economic, environmental, social, political, ethical, health and safety, manufacturability, and sustainability
(d) an ability to function on multidisciplinary teams
(e) an ability to identify, formulate, and solve engineering problems
(f) an understanding of professional and ethical responsibility
(g) an ability to communicate effectively
(h) the broad education necessary to understand the impact of engineering solutions in a global, economic, environmental, and societal context
(i) a recognition of the need for, and an ability to engage in life-long learning
(j) a knowledge of contemporary issues
(k) an ability to use the techniques, skills, and modern engineering tools necessary for engineering practice

The capstone design course serves an important role in graduate and undergraduate engineering programs. The capstone (or senior design) course covers in various degrees, all 11 of the ABET engineering program outcomes. The course provides engineering students open-ended project experiences with a variety of realistic requirements and constraints they might encounter in a rigorous project environment in industry. The teams learn by doing and reflection. Faculty mentors provide guidance and serve as advisors and coaches. The in-class discussions range from business entrepreneurship issues to technical dialogues. During class sessions, students have opportunities to learn about substantive engineering disciplines as well as reflect on the demands of professional practice.

The team accomplishes their work independently and collaboratively, primarily outside of the classroom. Class time is used for status review. The class that I mentor meets once a week. During the class we review each team's progress and gently (and sometimes not so gently) critique the team's work by asking penetrating questions about their progress and direction. The class sessions motivate the teams, if only by placing stress on the team members to show that the past week has been productive. During the class meeting, all class members may offer technical and nontechnical comments and suggestions that clarify and may expand or decrease the scope of a team's work. The constructive comments may recommend that the team meet with vendors, other professors, or experts from industry. Any topic dealing with the project is fair game including technical performance, budget and schedule questions, customer needs, test methods, and risk issues.

The class meeting simulates the weekly meeting usually held in industry between a project manager and technical project leaders. From the faculty mentor's perspective the goal of the weekly meetings goal is to make sure teams have a firm direction and actionable information, while making participants feel motivated and respected.

ABET accredits over 3,100 applied science, computing, engineering, and technology programs at more than 660 institutions in 23 nations. The engineering program accreditation criterion 5 [4] requires that engineering programs include a capstone project experience in the curriculum. Criterion 5 states the following:

> Students must be prepared for engineering practice through a curriculum culminating in a major design experience based on the knowledge and skills acquired in earlier course work and incorporating appropriate engineering standards and multiple realistic constraints.

The capstone course approach described in this book meets the intent of ABET general criteria 3 and 5.

Engineers Canada, through the Canadian Engineering Accreditation Board, accredits undergraduate engineering programs at Canadian higher education institutions. Students graduating from Canadian accredited engineering programs must possess attributes in the following categories [5]:

- **A knowledge base for engineering**: Demonstrated competence in university-level mathematics, natural sciences, engineering fundamentals, and specialized engineering knowledge appropriate to the program.
- **Problem analysis**: An ability to use appropriate knowledge and skills to identify, formulate, analyze, and solve complex engineering problems in order to reach substantiated conclusions.
- **Investigation**: An ability to conduct investigations of complex problems by methods that include appropriate experiments, analysis and interpretation of data, and synthesis of information in order to reach valid conclusions.
- **Design**: An ability to design solutions for complex, open-ended engineering problems and to design systems, components, or processes that meet specified needs with appropriate attention to health and safety risks, applicable standards, and economic, environmental, cultural, and societal considerations.
- **Use of engineering tools**: An ability to create, select, apply, adapt, and extend appropriate techniques, resources, and modern engineering tools to a range of engineering activities, from simple to complex, with an understanding of the associated limitations.
- **Individual and team work**: An ability to work effectively as a member and leader in teams, preferably in a multidisciplinary setting.
- **Communication skills**: An ability to communicate complex engineering concepts within the profession and with society at large. Such ability includes reading, writing, speaking and listening, and the ability to comprehend and write effective reports and design documentation, and to give and effectively respond to clear instructions.
- **Professionalism**: An understanding of the roles and responsibilities of the professional engineer in society, especially the primary role of protection of the public and the public interest.
- **Impact of engineering on society and the environment**: An ability to analyze social and environmental aspects of engineering activities. Such ability includes an understanding of the interactions that engineering has with the economic, social, health, safety, legal, and cultural aspects of society, the uncertainties in

the prediction of such interactions; and the concepts of sustainable design and development and environmental stewardship.
- **Ethics and equity**: An ability to apply professional ethics, accountability, and equity.
- **Economics and project management**: An ability to appropriately incorporate economics and business practices including project, risk, and change management into the practice of engineering and to understand their limitations.
- **Life-long learning**: An ability to identify and to address their own educational needs in a changing world in ways sufficient to maintain their competence and to allow them to contribute to the advancement of knowledge.

To help meet this broad list of attributes, many Canadian universities that offer engineering degrees include a team-based technical design project in their final year. As in U.S. engineering schools, the course begins with an understanding of the engineering design requirements. The courses' expected learning outcome is based on the knowledge and skills acquired by the student in earlier and concurrent courses. Canadian colleges encourage projects that serve societal needs and involve multidisciplinary elements.

Here again, the capstone course approach described in this book meets the intent of the attributes expected of students completing a degree in Canadian engineering schools.

The Springer Web site contains many of the templates discussed in this book. Please use them in your capstone class preparations.

Fairfield, CT, USA   Harvey F. Hoffman

Acknowledgements

I am grateful to all the professionals, professors, reviewers, and students who have provided valuable ideas and comments during the development of this project. I would especially like to thank and acknowledge the time and effort provided by Michael Blake, Esq. (Michael A. Blake, LLC, Registered Patent Attorney, Milford, CT and San Diego, CA) and John Yankovich, Esq. (partner in the firm Ohlandt, Greeley, Ruggiero & Perle, LLP in Stamford, CT) for reviewing the chapter on intellectual property.

Contents

1	**Engineering and the Capstone Course**	1
	Chapter Objectives	1
	What Is Engineering?	1
	The Capstone Course	2
2	**The Capstone Team**	7
	Chapter Objectives	7
	The Team	8
	Traditional Teams in Industry	8
	Tiger Teams	9
	Skunk Works	10
	Capstone Project Team	12
	Steps in Team Formation	13
	Team Conflict Management	15
	Common Causes of Conflict	15
	Cultural Styles and Conflict	19
	Ethics in Project Teams	21
	Project Team Peer and Self Evaluation	22
3	**Basic Team Communications**	25
	Chapter Objectives	25
	Conducting meetings	26
	Meeting Preparation	26
	After the Meeting	27
	Thinking About the Project Topic	28
	Brainstorming	28
	Presentations	30
	Preparing for the Presentation	30
	Powerpoint Presentations	31
4	**Capstone Class Written and Oral Submittals**	39
	Chapter Objectives	39
	Project Selection	42
	Topic Selection: Search Phase	42
	Weekly Progress Status Presentation: Project Search Phase	47

	Proposal Preparation and Presentations...	48
	Initial Activities..	48
	Weekly Progress Status Presentation: Project Proposal Phase	54
	Preliminary Design Review (PDR) and Critical Design Review (CDR)..	65
	Project Specification	66
	Proposal Report..	71
	Proposal Presentation...	81
	Project Development..	82
	Weekly Progress Status Presentation: Project Development Progress Phase	82
	Final Report, Presentation, and Demonstration	83
	Final Project Report	83
	Final Presentation...	85
5	**Project Development**...	87
	Chapter Objectives...	87
	The Engineering Design Process	88
	The NASA Design Approach	89
	Design Verification and Validation	91
	Design Verification Plan	91
	DRIDS-V Design Approach and Plan	95
6	**Innovative Capstone Project Examples**	99
	Chapter Objectives...	99
	Product or System Lifecycle	99
	Common Capstone Team Issues	102
	Capstone Project Examples...	102
	Web-Based Game...	103
	Town Intranet Portal...	104
	International Student Transition System...............................	104
	Hot Water Heater Information Tool	104
	Biometric Signature Solution for a State Medical Facility	104
	Emergency Operations Center (EOC) Design and Implementation	104
	Parking Solutions	105
	Phone-2-Phone (P-2-P) Translator Application for a Smart Phone...	105
	Chimney Sweep	105
	Green Fertilizer	105
	Rainwater Harvesting System Project....................................	105
	Helmet Integrated Impact Detection System	106
	Fitness Monitoring Device with Smartphone Application and Conventional Bluetooth...	106
	Voice-Controlled Wheelchair..	106
	Snow Melt Mat..	106

Snow Load Sensor Mat	107
Mini-Hydro Potable Water System	107
Fire Hydrant Locator	107
Firefighter Training Vent Simulator	107
Parts Protection Bag	108
Medical Reminder System (MRS)	108
PALS: Personal Alert Location System	108
High Wind Speed Alternate Energy System	108
HeDS-UP: Motorcycle HeDS-UP (Helmet Display System)	108
Automatic Ladder Leveler	109
Solar Powered Food Dehydrator	109
The Elements of a Successful Project	109
7 Intellectual Property	**113**
Chapter Objectives	113
Intellectual Property	114
Patent Search	115
Example Patent Search	116
Types of Patents	121
8 Epilogue	**123**
A Fable	124
A Perspective	124
Appendix: Failure Mode and Effects Analysis (FMEA) and House of Quality	**127**
References	**139**
Index	**143**

List of Figures

Fig. 1.1	Capstone project timeline	4
Fig. 3.1	Presentation do's and don'ts	34
Fig. 4.1	Risk matrix	62
Fig. 4.2	Converting the risk matrix to a severity priority list	63
Fig. 5.1	Complete design process	90
Fig. 5.2	Product, service, or system verification process	92
Fig. 6.1	Product or system lifecycle	100
Fig. 6.2	Elements for project success	110
Fig. 7.1	The United States Patent and Trademark Office website	117
Fig. 7.2	Searching the USPTO database	118
Fig. 7.3	A list of patents containing the words sandbag, salt, polymer	119
Fig. 7.4	One patent containing the words sandbag, salt, polymer	120

List of Tables

Table 2.1	Peer and team evaluation instrument	23
Table 3.1	Project meeting agenda	27
Table 3.2	Project meeting minutes template	28
Table 4.1	Typical SWOT analysis factors for a capstone team's idea	46
Table 4.2	Product comparison template	49
Table 4.3	Typical capstone project proposal tasks	51
Table 4.4	Typical capstone project development tasks	51
Table 4.5	Capstone project budget template	52
Table 4.6	Capstone project weekly progress status log	55
Table 4.7	Capstone project budget status update—variance analysis	57
Table 4.8	Typical capstone project proposal work breakdown structure (WBS) and schedule	59
Table 4.9	Risk matrix—probability score	63
Table 4.10	Risk matrix—impact score	63
Table 4.11	Severity calculation	64
Table 4.12	Risk severity score	64
Table 4.13	Criteria for a good requirement	67
Table 4.14	List of project deliverables	76
Table 4.15	Multiyear business revenue and cost work sheet	78
Table 5.1	The cost of correcting engineering design errors during project phases	89
Table 5.2	Comparison between design verification and validation	91
Table 5.3	Example product design verification matrix	93
Table 6.1	Common 'ilities attributes	101
Table 7.1	Provisional/Nonprovisional patent application comparison	122

Appendix
Table 1	Failure mode and effects analysis (FMEA) worksheet	129
Table 2	Failure mode and effects analysis (FMEA) worksheet	130
Table 3	FMEA severity	131
Table 4	FMEA probability of occurrence	131
Table 5	FMEA detectability	132
Table 6	House of quality	133
Table 7	House of quality for a road bike drop handlebar	136

About the Author

Dr. Harvey F. Hoffman's background includes both industry and academic experiences. While in industry, he held positions ranging from research and development engineer, group manager, project manager to general manager. In the aerospace field he worked on the space shuttle and the Titan rocket. He participated in the development and production of radar and artillery weapon systems; and analytical instruments, systems, and services for radiation detection and radiation monitoring. He holds a patent for a Digital Scan Converter Employing an Accelerated Scan. Dr. Hoffman earned five degrees including a master's degree in management, a doctorate in higher educational administration and three degrees in electrical engineering. He has been an accreditation evaluator for the Accreditation Board for Engineering and Technology (ABET), the New York State Department of Education and the Accrediting Council for Independent Colleges and Schools (ACICS). He is a past member of the Board of Examiners for the Malcolm Baldrige National Quality Award (MBNQA).

In academe, Dr. Hoffman has taught both graduate and undergraduate courses at the New York Institute of Technology, Bridgeport Engineering Institute (BEI) and Fairfield University. He served as the Chair of the Electrical Engineering Department at the Bridgeport Engineering Institute and Fairfield University. While at the Bridgeport Engineering Institute, Dr. Hoffman initiated and led the introduction of a baccalaureate in Information Systems Engineering, which was the forerunner of the present Fairfield University B.S. in Computer Engineering. At BEI, he served as the first department chair for Information Systems Engineering and led a committee that performed the preliminary development of an electrical engineering master's degree program.

His career as an academic administrator includes serving as Dean and later as Vice President for Academic Affairs at TCI in New York City, the Provost at Lehigh Valley College in Pennsylvania, and the Vice President for Institutional Effectiveness at ASA Institute in New York City.

He is presently director of the Management of Technology graduate degree program at Fairfield University. He developed and teaches the Leadership, Project Management and Quality Management courses in the School of Engineering at Fairfield University. He also teaches Finance and Quality Management in Healthcare in the School of Nursing.

Dr. Hoffman has extensive experience on local and national Boards and Committees. He has served on, among others, the Computer Technology Industry Association (CompTIA) Project+Cornerstone Committee, the Board of Directors of the Norden Systems Federal Credit Union, the United Technologies Technical Education and Training Advisory Council, and the Lehigh Valley Executive Forum.

Dr. Hoffman has been the recipient of several awards including membership in the honor societies Sigma Xi, Eta Kappa Nu, and Phi Delta Kappa. He received the 1987 UTC Horner Citation for technical contributions to the Titan 34D program. In academics, he received the 1988-1989 Educator of the Year at the Bridgeport Engineering Institute and was elected a Fellow of the Institute in 1991. He was awarded the TCI Outstanding Service Award in 2001.

He is the author of a project management textbook (*Organizations Through the Eyes of a Project Manager*) and more than 20 academic and technical papers and presentations.

Engineering and the Capstone Course

> *Scientists discover the world that exists; engineers create the world that never was.*
> —Theodore Von Karman, Aerospace engineer
>
> *Scientists dream about doing great things. Engineers do them.*
> —James A. Michener, Novelist

Chapter Objectives

After studying this chapter, you should be able to:
Understand the field of engineering.
Describe what engineers do.
Gain familiarity with the basic elements of a capstone or project design course.
Explain the capstone or senior project timeline.

What Is Engineering?

No profession unleashes the spirit of innovation like engineering. From research to real-world applications, engineers constantly discover how to improve our lives by creating bold new solutions that connect science to life in unexpected, forward-thinking ways. Few professions turn so many ideas into so many realities. Few have such a direct and positive effect on people's everyday lives. We are counting on engineers and their imaginations to help meet the needs of the twenty-first century [6].

Engineering is about making products and services—making things better, safer, more reliable, with improved quality and performance—that meet a specification, and completing the project within a prescribed budget and schedule, while keeping a customer happy. Engineering is a pragmatic profession. Engineers are builders, entrepreneurs, and problem solvers. They seek quicker, better, more-efficient, and less-expensive ways of completing projects. They seek to do things that have never been done before or improve a product or service that has been around for a long time. Depending on their specialty, engineers get involved with product, service, process and system design, development, research, manufacturing, test, marketing, and sales. They create hardware and software products and services. They develop and design products that make use of chemical, physical, or biological transformations of raw materials. They deal with new technology but also deal with the complexity of integrating new technology into legacy systems. They must consider and understand the business, legal, and societal aspects of their work.

Engineers make a difference by turning ideas into reality. Engineers help meet many societal needs in fields like—housing; computers and computing systems; defense; aerospace; avionics; medical devices and systems; telecommunications devices and systems; air, ground, and sea transportation; ecology; energy production and power generation and distribution. They develop new products and services, and translate technical principles into goods, processes, and systems for humanity's benefit.

Engineers work on activities involving an actual or anticipated qualitative or quantitative need. Organizations like Engineers Without Borders implement sustainable engineering solutions for water supply, sanitation, energy, agriculture, civil works, structures, and information systems projects. Engineers serve as government liaisons to assist new and existing organizations seeking assistance with technology related issues. Many engineers work not only to develop new products and services but also to manage production operations and help maximize a company's profitability by increasing revenue and lowering operating expenses.

Engineers rarely work alone. They most often work as part of a team. Indeed, engineers cannot work in a vacuum. Their work requires them to interact with customers, suppliers, vendors, manufacturers, sales personnel, management personnel, and other engineers on a regular basis.

The Capstone Course

A capstone course is intended to be the final course in a major. It makes use of prior coursework and life experience in a cumulative and integrative fashion. Capstone courses exist in virtually all academic disciplines. An engineering capstone course integrates the skills and competencies that students have learned in their engineering program. It attempts to balance technical, business, and interpersonal skills that will help students to immediately contribute to team efforts in today's fast-paced business and technical environment. The engineering capstone course simulates as close as academically possible the activities in which an engineer is involved. The course challenges the student's personal and professional skills. The nature of the course forces us to accept dimensions of professional practice that go beyond technology by also including societal considerations. Most engineering capstone courses expect participants to apply the information gained from engineering, English writing, economics, statistics, accounting, management, marketing, law, ethics, and computer engineering to the task-at-hand.

Many authors (see for example [7–11]) have discussed the purpose of the capstone course at the undergraduate and graduate levels. These authors agree that a capstone course should encompass technical and social practices. Specifically, the capstone course should include experiences that benefit the graduate or senior undergraduate student learner by:
- Using technical knowledge towards solving an engineering problem that incorporates "real world" practical issues and challenges.
- Integrating technical and non-technical competencies.
- Using management techniques to mitigate and overcome problems and issues that invariably arise during a project.
- Engaging in a team activity that will call upon the student's to use collaboration, conflict resolution, and brainstorming skills.
- Practicing oral and written communication skills.

Capstone courses should challenge students and take them out of their comfort zones. They are intended to provide a simulated "real-world" start-to-finish learning experience. They will learn to anticipate, plan, and manage change in development projects. The ten tasks involved in a typical capstone project class are:
1. Identifying a need for a product, service, process, or system.
2. Forming a team.
3. Identifying competitive products and services, performing a patent search, and identifying required resources.
4. Refining the topic and generating a project proposal that includes a specification, a task list or a work breakdown, and a schedule.
5. Preparing a design.
6. Developing and fabricating the product, process, or service.
7. Developing a test plan that complies with the specification.
8. Testing the system, product, process, or service.
9. Examining the business viability aspects of the project.
10. Preparing a final report and making a presentation including a demonstration of the product, process, or service.

Figure 1.1 summarizes the capstone or senior project timeline. Note that the first item to consider is need or benefit to the organization or society. The benefits anticipated as a result of the project should be identified and defined in as measurable terms as possible. A business case to implement a project begins with a need. The team must also consider a way to quantify that benefit. The following are some needs that may stimulate thinking about the benefits of a project idea:
- To comply with new legislation.
- To provide a more efficient and effective operation.
- To improve the organization's ecological participation.
- To provide a simpler, newer, or better service to the public.
- To improve an existing product, service, or process.
- To reduce existing costs.
- To avoid or reduce future costs.
- To modernize the working environment and conditions for employees.
- To improve communication within and beyond the organization.
- To reduce the amount of effort required to follow up mistakes and complaints.

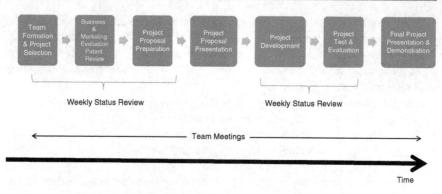

Fig. 1.1 Capstone project timeline

- To improve the quality of information and decision making.
- To take advantage of new technology.
- To extend the operation and life of legacy systems.
- To increase morale and motivation.
- To enable other initiatives to deliver benefits to the organization.

Throughout this process the team reviews their proposed solution with regard to meeting the original specification and assesses risks that arise that require project modifications while keeping the schedule and budget in mind. At times, a significant change in direction is required. When making a change remember the team adjusts their plans to reflect

1. The tasks to be added to accomplish the change, including requirement gathering, changes to design, additional purchases, additional testing, and so on.
2. New resources and skill sets that may be required.
3. The schedule impact.
4. The cost impact.
5. New risks introduced by the changes.

Ultimately, the project and the project team will be evaluated based on the team's outcomes. The capstone course helps students to critically think about a project. Student's refine their ability to identify, analyze, and solve technical problems. They apply their creativity to the design of systems, products, services, components, and/or processes. To assist in the process, they use software and hardware tools to which they have been exposed during their course of studies. They may have to learn to use a tool for which they did not have prior experience. In addition to applying standard engineering methods and techniques to carrying out the Project, they plan the project using management tools such as Microsoft Project, Open Project (https://www.open-project.org/), or Primavera, which is used to develop a task schedule. Students learn the importance of effective oral and written communication with their team, mentor, advisers, and vendors. The team quickly recognizes the need for timeliness in meeting their assigned tasks. During their time together, the team becomes sensitive to the importance of meeting their professional and ethical responsibilities to one another.

A capstone course is useful to an institution as part of its assessment practices. A capstone course assists the institution in demonstrating that the student has achieved some or all of the eleven outcomes established by the Accreditation Board for Engineering and Technology (ABET) for graduate and undergraduate programs [4].

Students gain a sense of satisfaction in completing a comprehensive project as a team member while using all their newly developed skills. While there are rules to the game, the team is primarily acting on its own. Students feel a sense of ownership, they soon forget that this is a required academic task and genuinely feel the pure enjoyment of learning and accomplishment.

Finally, a motivational thought about the capstone team project. If by the end of the semester the team thinks that it has a viable product, it might consider contacting a firm like Quirky (http://www.quirky.com/). Quirky is a New York-based startup that designs and builds consumer products created and refined by its community of users. Quirky solicits ideas for inventions that its users can put through a design process. The company then gets finished products placed on the shelves of major retailers. In 2013, GE arranged a deal with Quirky to build 30 connected-home devices in the next 5 years [12]. So, if the team comes up with an interesting idea during the capstone journey, consider joining a community of innovators and bring it to the market.

The word stakeholder is used throughout the book. Stakeholder is in some sense a nebulous term. Think of it as a person, group, or organization that has an interest in the project. The primary stakeholders in a typical corporation are its investors, employees, customers, and suppliers. Under some circumstances, the community, government, and trade associations may also have an interest in a project and may be important stakeholders. The stakeholders in the capstone course are students, of course, teachers, and college administrators. Sometimes community members have a vested interest in the success of a project, as does a student's employer.

2. The Capstone Team

> *The engineer is the key figure in the material progress of the world. It is his engineering that makes a reality of the potential value of science by translating scientific knowledge into tools, resources, energy and labor to bring them into the service of man ... To make contributions of this kind the engineer requires the imagination to visualize the needs of society and to appreciate what is possible as well as the technological and broad social age understanding to bring his vision to reality.*
>
> —Sir Eric Ashby

> *Engineering is the science of economy, of conserving the energy, kinetic and potential, provided and stored up by nature for the use of man. It is the business of engineering to utilize this energy to the best advantage, so that there may be the least possible waste.*
>
> —William A. Smith

Chapter Objectives

After studying this chapter, you should be able to:
Understand what a team is.
Compare different types of teams.
Gain insight into the elements of forming a team.
Explain team development and behavior.
Understand the basics of conflict management associated with teams.
Relate ethics to project teams.
Understand how to assess team performance.

The Team

> A team is a small group of people with complementary skills who are committed to a common purpose, performance goals, and approach for which they hold themselves mutually accountable. [13]

Teams are ubiquitous in industry today. Alexander Graham Bell's and Thomas Alva Edison's experience of working in comparative isolation is gone. Today's engineering projects require technically competent engineers working as team members with other professionals. It is the ability to work in multidisciplinary teams communicating and sharing information that enables them to complete a project on time, within budget, and meeting the intended requirements.

Teams are a part of business and industry. Some teams exist for a longer period and may be considered permanent. Most teams exist to complete a task or project. Researchers [13] write that there are three distinct types of teams:
1. Teams that recommend things.
2. Teams that make or do things.
3. Teams that run things.

Teams are made of individuals whose collective competence and experience is greater than any individual offers. Teams must be planned, supported, and led. Members interact regularly, coordinate work efforts, and engage in healthy conflict. Team members listen to one another and learn to respect each other's opinion even if they disagree. They develop a feeling of loyalty and togetherness.

Traditional Teams in Industry

Most technology companies organize their employees by function, such as engineering, project management, technical writing, production, field service, finance, sales, and marketing. The project team in a standard matrix organization requires a project manager to request needed personnel from each functional group. Each functional manager contributes personnel to the project team. Led by the project manager, the newly formed team gets together and collaborates to meet a goal, such as developing a new product. Each project member adds to the overall project success by bringing his or her specialty skills and competencies to the team. As in most projects, the team receives or creates a specification, a budget, and a schedule. They may be colocated or work from their functional areas. The team is expected to follow established processes and procedures for purchasing material and services, testing, standards usage, quality methods, documentation, training, upper management supervision, etc. They adhere precisely to the organization's policies and procedures related to personnel issues—especially those associated with union shops.

The New United Motor Manufacturing Inc. (NUMMI) plant in Fremont, Calif., opened in 1984 as a joint venture between General Motors (GM) and Toyota. It represented an opportunity for GM to gain technology and insights into Toyota's production system, and Toyota would learn how to apply its systems and culture to a U.S. workforce. The plant operated until 2010. "Team" was at the heart of the production system. The "Toyota Way" demanded trust and respect in worker teams. They implemented flexible work rules and empowered workers to stop the production line if a problem needed to be fixed; rewarding workers for solving problems in production and improving product quality. (As a footnote, Tesla bought the facility from Toyota and began to produce the Tesla Model S electric sedan in 2012 at the former NUMMI plant.)

Tiger Teams

Unlike traditional teams, skunk works and tiger teams are frequently assembled to deal with special situations that may arise in business and industry. The Tiger Team enters the picture when a crisis emerges. Management quickly assembles a multi-disciplinary team of experts for the purpose of problem solving.

Tiger Teams have been used in aerospace, computer security, and the military. Tiger teams may test an organization's security measures to see how easily they could be penetrated. A 1964 definition of Tiger Teams described them as "a team of undomesticated and uninhibited technical specialists, selected for their experience, energy, and imagination, and assigned to track down relentlessly every possible source of failure in a spacecraft subsystem" [14]. A corporate sponsor must provide the needed resources, especially budget and personnel. The tiger team is a self-contained crew that includes all the skill sets and resources needed to do the assigned work. Team members could be drawn from engineering, operations, finance, legal, and marketing areas. Consultants frequently supplement internal personnel to accomplish a task.

An organization does not use tiger teams for every issue that arises. It costs too much. The tiger team represents the best that any group, organization, or government can assemble, to attack a daunting problem. Characteristics in the performance of a Tiger Team are [15]:

- The ability to arrive at multi-dimensional solutions at three levels: technical, process, and human
- The ability to incorporate statistical and scientific methods of problem solving and decision making
- A willingness to break rules, think outside the box, and move beyond existing boundaries
- The ability to maintain a continuous intensity of focus and action orientation from all members, not just the leader, over the entire span of the work
- The capability of addressing complex, multi-faceted tasks and/or projects with narrow margins for error
- The ability to perform within tight timeframes and low risk tolerances to achieve rapid response recovery

Perhaps one of the most well-known tiger teams was the 15 member team to get the Apollo 13 flight in 1970 on the correct path home. NASA lead flight director Gene Kranz told them to solve the problem [16]. As with all tiger teams faced with adversity and almost insurmountable problems, the leader must create a climate of trust, cooperation, and innovation, and lead with "Can do!" optimism and enthusiasm.

Skunk Works

Skunk works involves a multidisciplinary team assigned to a special project, usually working with advanced technology, limited budgets, and aggressive schedules. The projects are designed and built relatively quickly with minimum management constraints.

During World War 2, America needed an aircraft that could meet and exceed the capabilities of German jets. In 1943, the U.S. War Department hired Lockheed Aircraft to build a working jet fighter prototype in less than six months. A team lead by Kelly Johnson moved forward with minimal bureaucracy and designed and built the aircraft within 143 days [17]. The Skunk Works® name is a Lockheed Martin registered trademark. Johnson used an unconventional organizational management approach. He broke the rules and challenged the system in his effort to improve efficiency and obtain results. His philosophy is spelled out in his "14 rules and practices" [18], which are:

1. The Skunk Works manager must be delegated practically complete control of his program in all aspects. He should report to a division president or higher.
2. Strong but small project offices must be provided by both the military and industry.
3. The number of people having any connection with the project must be restricted in an almost vicious manner. Use a small number of good people (10–25 % compared to the so-called normal systems).
4. A very simple drawing and drawing release system with great flexibility for making changes must be provided.
5. There must be a minimum number of reports required, but important work must be recorded thoroughly.
6. There must be a monthly cost review covering not only what has been spent and committed but also projected costs to the conclusion of the program.
7. The contractor must be delegated and must assume more than normal responsibility to get good vendor bids for subcontract on the project. Commercial bid procedures are very often better than military ones.
8. The inspection system as currently used by the Skunk Works, which has been approved by both the Air Force and Navy, meets the intent of existing military requirements and should be used on new projects. Push more basic inspection responsibility back to subcontractors and vendors. Don't duplicate so much inspection.
9. The contractor must be delegated the authority to test his final product in flight. He can and must test it in the initial stages. If he doesn't, he rapidly loses his competency to design other vehicles.
10. The specifications applying to the hardware must be agreed to well in advance of contracting. The Skunk Works practice of having a specification section stat-

ing clearly which important military specification items will not knowingly be complied with and reasons therefore is highly recommended.
11. Funding a program must be timely so that the contractor doesn't have to keep running to the bank to support government projects.
12. There must be mutual trust between the military project organization and the contractor, the very close cooperation and liaison on a day-to-day basis. This cuts down misunderstanding and correspondence to an absolute minimum.
13. Access by outsiders to the project and its personnel must be strictly controlled by appropriate security measures.
14. Because only a few people will be used in engineering and most other areas, ways must be provided to reward good performance by pay not based on the number of personnel supervised.

The skunk works legacy has continued and has made business history [19]. Organizations use the skunk works-like project process where teams operate independently from the structures of the remainder of the organization in an effort to spark innovation and complete a task in a shorter than traditional time period.

The Motorola Razr phone is a case in point. The team kept the project top-secret, even from their colleagues. They used materials and techniques Motorola had never tried before. They threw out accepted models of what a mobile telephone should look and feel like. The team designed and constructed a phone with the features they wanted—including size. The team decided on a phone that was 13.9 mm thick, which was 40 % thinner than Motorola's slimmest flip-top phones. The 20-person engineering team completed the project in about a year and a half [20].

The Apple Macintosh computer resulted from an initial team of four, headed by Apple cofounder Steve Jobs. Their goal was to make a personal computer easy enough for an ordinary person to use without fear and inexpensive enough to be affordable. Secretly and in a separate facility, the Mac team took 3 years to develop the computer. It featured an intuitive graphic user interface that allowed nonprogrammers to use it almost instantly.

The development of the Ford (F) diesel engine Scorpion was developed in 36 months versus the more common 48 months [21]. Here again, the assembled team moved off-site to "short-circuit" the usual development process.

IBM's first PC was assigned the code name "Acorn." A small skunk works team of engineers, worked at a site in Boca Raton, Florida, to design and build it [22]. On August 12, 1981, after about a year of work, IBM released their new computer with the new name—the IBM PC (personal computer).

The Lockheed Skunk Works process demonstrated a method to rapidly prototype, develop, and produce a wide range of advanced aircraft for the U.S. military. The skunk works process continues and is now synonymous with projects that are designed and built quickly and unconventionally with minimum management constraints.

The preparation of proposal responses to government and private industry procurements is frequently managed as a skunk works effort. A proposal manager has a relatively short timeframe to provide a response. Subject matter experts are quickly brought on board to provide the specialized expertise that supports the development of the technical approach and related writing. The team has the responsibility for meeting schedule deadlines, conforming to the proposal outline and compliance matrix, developing graphics that support the text, developing management and cost volumes while using a consistent format. A contracts specialist is generally responsible for preparing required forms, clauses, representations, and certifications, which are included in the cost/pricing section. The team members must cooperate in a coordinated effort to understand the client's goals and anticipate the competitors' approaches. A winning proposal team has to do whatever it takes over the typically allotted 30- to 90-day proposal preparatory time period. In this way it is very similar to a skunk works operation.

Capstone Project Team

A capstone project team frequently consists of from 2 to 5 people. Students sometimes take the initiative to form their team in a self-selection process. Other times an instructor or mentor will either randomly assign students to a team or assist students in forming a team.

The capstone team exists for one or two semesters and its members share a common purpose throughout this time. Whether self-selected or pre-selected, the members should be a diverse group and have a broad collective skill and knowledge base. A team composed of like-thinking individuals with similar backgrounds and experience may limit the number and types of possible solutions for creative problem solving.

In many respects, the capstone project team is like a skunk works. Not all of the Kelly Johnson rules apply to the capstone project team, but certainly a subset does. Consider adopting the following set of modified skunk works ground rules for the capstone team:

1. The capstone project team has almost complete control of their project. They report only to the class faculty mentor.
2. The number of people having a connection with the project is restricted to the capstone project team and the project mentor. The capstone team has the option of calling in industry or university consultants for guidance.
3. The capstone project team has great flexibility for making changes but all participants must be notified of a change.
4. The capstone project team has a minimum number of required reports, but important work must be recorded thoroughly. The required reports include a weekly status review, a final semester presentation, and a summary semester report.
5. The capstone project team must set, commit to, and meet milestones to keep momentum going.
6. The capstone project team must keep close tabs on expenditures. The weekly status report should include a cost review covering not only what has been spent and committed but also projected costs.

7. The capstone project team is expected to obtain good vendor bids for material and subcontract work used on the project.
8. The capstone project team must have a well-defined specification before beginning significant work on the project.
9. The capstone project team must develop a test that demonstrates that the final product or service meets the specification.
10. There must be mutual trust and cooperation among the members of the capstone project team and the faculty mentor.
11. The team must meet frequently, distribute responsibility as evenly as possible and offer one another constructive feedback.
12. Have fun and celebrate success!

The team develops performance goals that will enable them to establish, track, and evaluate progress towards completing their project. Teams develop their own rules that outline the expected behaviors of its members including when and how often to meet. One of the most challenging tasks that a team faces is finding meeting times outside of the classroom that are acceptable to all. This is especially true of part-time students working during the day and having availability on particular evenings during the week. Indeed, if the student has children, it quickly becomes a family discussion item that must be resolved. At the end of the process, the team will own and share the team's outcomes—both successes and failures.

Steps in Team Formation

Members of a healthy team encourage listening and respond constructively to views expressed by others. They provide support and recognize the interests and achievements of others. A team's performance includes both individual results and "collective work products" [13]. A collective work product is the result of an output that members work on together, such as documentation, subcontracting, or test results. A collective work product reflects the joint, real contribution of team members.

Bruce Tuckman [23] proposed a 4-stage model of group development, which is applicable to the capstone team process. Broadly, the specific features of each stage are:
1. Forming: The team comes together and gets to know one another; form as a team and begin to understand the task ahead. The team looks to the faculty mentor and other team members for guidance and direction. Initially, little agreement exists on the team's aims. Individual roles and responsibilities are unclear. The faculty mentor responds to questions about the team's purpose, objectives, and external relationships. Processes may be ignored. Rules of behavior are to keep things simple and to avoid controversy. Team members may question the need for the course, and test the system and mentor.

   Team members attempt to become oriented to the tasks as well as to one another. Team discussions center around defining the task scope, how to approach it, and other similar concerns. To grow from this stage to the next, each member must relinquish the comfort of nonthreatening topics and risk the possibility of conflict.

2. Storming: Storming is characterized by competition and conflict in the personal relationships and in the task assignments. Decisions may not come easily. Team members vie for position as they attempt to establish themselves in relation to other team members, the initial team leader, and faculty mentor. Clarity of purpose increases but uncertainties persist. Factions form and there may be power struggles. Some team members may be dissatisfied with the way that work has been distributed. The team needs to focus on its goals to avoid becoming distracted by relationships and emotional issues. Compromise is required to enable progress.

    As the team members attempt to organize, conflict may arise. Individuals have to bend and mold their feelings, ideas, attitudes, and beliefs to suit the team. Although conflicts may or may not surface, they will exist. Members may voice dissatisfaction about the project selected or the tasks assigned to them. Questions will arise about who is going to be responsible for what, what the schedule is, and what the internal evaluation criteria are. There may be conflict over leadership, structure, power, and authority. There may be wide swings in members' behavior based on the stress associated with emerging issues. Because of the discomfort generated during this stage, some members may remain completely silent while others attempt to dominate.

    In order to progress to the next stage, group members must move towards a problem-solving mentality. Important traits in helping teams move on to the next stage may be their abilities to listen, negotiate, and conciliate.

3. Norming: Eventually agreement is reached on how the team operates. Team members acknowledge all members' contributions. Members are willing to change their preconceived ideas or opinions on the basis of facts presented by other members, and they actively ask questions of one another. Leadership may be shared, and factions dissolve. When members begin to know—and identify with—one another, the level of trust in their personal relations contributes to the development of group cohesion. The "*I*" disappears and the "we" appears. It is during this stage of development (assuming the group gets this far) that people begin to experience a sense of team belonging as a result of resolving conflicts.

    Team members share information, feelings, and ideas. They solicit and give feedback to one another, and explore actions related to the task. During the norming stage, team members' interactions are characterized by openness and sharing of information on both a personal and task level. They feel good about being part of an effective group.

    However, be careful there can be a downside to a cohesive group. Teams in the norming phase increase their commitment to the team. As cohesion increases, performance norms are established and members tend to want to increase conformity to the standards that are set. High conformity may incur groupthink.

    Groupthink [24] occurs when a group makes faulty decisions because group pressures lead to a deterioration of "mental efficiency and reality testing." Groups affected by groupthink ignore alternatives. A group is especially vulnerable to groupthink when its members have similar backgrounds, when the group is insulated from outside opinions, and when there are no clear rules for decision making.

The effects of groupthink may reduce innovation and effective decision making. The team may become uninspired to think independently or to consider ideas or solutions that run counter to those supported by the majority of the team.

4. Performing: The team practices and begins to get good at what it is doing. Participants become effective in meeting objectives. In this stage, people can work independently, in subgroups, or as a total unit with equal facility. Their roles and authorities adjust to the team's changing needs. The performing stage is marked by interdependence in personal relations and problem solving. The team is most productive. Individual members have become self-assured. Members are both task oriented and people oriented. A team identity exists, team morale is high, and team loyalty is strong. The team focuses on solving problems and emphasizes achievement.

Tuckman together with Jensen [25] added a fifth stage (adjourning) years later.

5. Adjourning: Projects end and so does the team. After the team has successfully (or unsuccessfully, in some cases) completed their task, they must disband. Participants disengage from the team and move on. Typically in the college capstone environment this would involve a final presentation with recognition for everyone's participation and achievement. In just a few short days after the presentation the team members graduate and say their personal goodbyes. Students graduate with a new degree and move on to face new challenges. They are confronted with the task of finding a job or moving to the next step in their current position. Concluding a team effort in industry can create some apprehension because the former team members are concerned about their next assignment.

The Tuckman model tells us that over time a team develops and grows in ability and trust. The members' behavior changes through the experience. While the team experiences ups and downs, the model does not predict when the team goes through the phases or even if the team will experience every phase. However, the science of organizational behavior suggests that the Tuckman model is representative of the process that your team will experience. Bonebright states that "It is, perhaps, unlikely that a model with similar impact will come out of the new literature." [26]

One thing is certain—the team's priority is to finish satisfactorily so that each member can obtain his or her degree. The team has no choice but to work together and succeed. In the process, members will learn a good deal about themselves, teamwork, and the developed product, service, or process.

Team Conflict Management

Common Causes of Conflict

Tuckman discussed the idea of conflict in the storming phase. Anticipate team conflict. People are different and putting people with different backgrounds, personalities, and experiences together will necessarily yield a variety of opinions, insights, and ideas. More often than not, the people's diversity leads to better decision making.

If everyone agreed completely with one another and knew the same information they might have little to contribute.

For an effective team that will reach its goals, the team members must have a shared understanding of what they are striving to achieve, as well as clear objectives. The team members need to keep personal conflict to a minimum. Personality conflict, may lead to a lack of cooperation, a lack of communication between members and unprofessional behavior that can directly affect the entire group. Team members must understand the rudiments of solving problems caused by conflict, before conflict becomes a major obstacle to completing its work.

Common causes of conflict within a team include the following:
- Disagreements in the technical approach
- Intolerance for mistakes
- Using alternate methods for accomplishing a task
- Lack of trust
- Different cultures, values, attitudes, languages, vocabularies, and perceptions
- Lack of meeting time
- Differences in objectives and different understandings of productive work
- Team members failing to meet their assigned work tasks in a timely fashion
- Scarcity of resources (finance, equipment, facilities, etc.)
- Disagreements about needs, goals, priorities, and interests
- Poor communication
- Lack of clarity in roles and responsibilities
- Hoarding rather than sharing knowledge

Other symptoms of team conflict also include:
- Gossip
- Not returning phone calls, texts, or e-mails
- Not responding to requests for information
- Hostility
- Excessive complaining
- Finger pointing
- Verbal abuse
- Not attending required meetings
- Absenteeism
- Physical violence
- Sexual harassment

People have different styles of communication, different political or religious views and different cultural backgrounds. In our diverse society, the possibility of these differences leading to conflict exists, and the team must be alert to prevent and resolve situations where conflict arises.

Teams formed in large university environments will likely be quite diverse with respect to age, gender, ethnic background, race, religion, language, and nationality. There may be significant cultural differences in the form of values, beliefs, attitudes, behaviors, and other intangibles that influence the team's interactions and may lead to nontechnical disagreements. Team members must put personality disagreements aside for the good of the project effort.

Conflict can be an effective means for everyone to grow, learn, and become more productive if it is resolved. Ongoing unresolved conflict may impede the team's efforts to complete the project. People frequently use one or more of the following options for managing conflict include the following:
- Avoidance—withdrawing from or ignoring conflict.
- Smoothing—playing down differences to ease conflict.
- Compromise—giving up something to gain something.
- Collaboration—mutual problem solving.
- Confrontation—verbalizing disagreements.
- Appeal to team objectives—highlighting the mutual need to reach a higher goal.
- Third-party intervention—asking an objective third party (project mentor) to mediate.

When team members think that progress is stalled or that the team members are not working well together, then meet and talk about it. The conflict resolution process begins by first acknowledging that there is a problem and moving towards defining the issue. Discussing a problem at an early stage can prevent small issues from escalating into major problems.

The team has to communicate in a clear and nonblaming manner. Gather data and separate fact from conjecture or assumptions. Make certain that every team member has the most recent information. Each team member should try to understand the other person's viewpoint. Attack problems not each other. Confirm that team members understand the team's goals and their individual roles.

Sometimes things don't go as planned. Treat a failure as an opportunity for team growth. In a nonthreatening way apply the "5 Whys" strategy. Discuss the problem and ask: "Why?" and "What caused this problem?" The answer to the first "why" will prompt another "why" and the answer to the second "why" will prompt another and so on. For example, if a Web site goes down ask:

Question	Possible response
Why was the Web site down?	The CPU utilization on all the front-end server went to 100 %
Why did the CPU usage increase to 100 %?	New code was added and it contained an infinite loop
Why did it contain an infinite loop?	The code was not completely tested
Why was the code not completely tested?	A new employee wrote the code and was not trained in software code test and verification methods

By posing the why question and examining the responses, we discover that the problem lies not so much with the newly developed code but with the organization's new employee training program. Appropriate corrective action can then be taken.

Repeating "why" several times helps to uncover the root cause of the problem and move towards correcting it. Although called the 5 Whys technique it does not necessarily have to be 5 Whys. It can for example be 4, 6, or 7. Other benefits of 5 Whys are the following:
- Easy to use and requires no advanced mathematics or tools.
- Separates symptoms from causes and identifies the root cause of a problem.
- Fosters teamwork.
- Inexpensive. It is a guided, team-focused exercise. There are no additional costs.

However, the 5 Whys method is not a perfect root cause approach to finding solutions to problems. It may not work well for complicated problems or problems with multiple causes.

When confronted with a technical or business issue that stumps the team, seek help. Review the availability of resources and whom team members might contact for guidance and support. Always start with the team's faculty mentor. In a large college or university there are a host of people with special skills of which the team could avail itself. Consider contacting adjunct faculty, traditional faculty, teaching and research assistants, laboratory assistants, and administrators. Be sure to consider contacting other departments if the need arises. Doing this will develop team member's relationship and political skills.

Hopefully, team members learn to overcome the deadly effects of procrastination. Every team member will likely tolerate some amount of inconvenience and delay—but each individual will have to set priorities in their personal lives to accommodate the overall team goals.

Many nontraditional students attend college while balancing the stresses of a family and a job. Work-study-life balance exacts huge demands and responsibilities on people's lives. Stress is an inevitable part of being a student. There is no silver bullet to use as a guide to assist a student to overcome the effects of stress. Plan and account for time—from the few hours blocked off for sleep, to the daily commute and lunch break.

Hopefully, by the time people have reached the capstone course he or she will have learned to balance the responsibilities associated with their personal life. Family members and friends will have learned to accept a person's unavailability some evenings and weekends. Nonetheless, there is considerable stress involved in telling a boss that you are not available for travel or that you can't work too many weekends. Be sure to inform your employer, friends, clients, and family about your schedule. The capstone class involves time in class, but the fact is that most of the work is done outside of the classroom—both independently and with the team. The good part is that it will be over in one or two terms. Then life will return to normalcy.

With appropriate planning, the team has the skills and abilities to perform the requisite tasks and it will not need to resort to obtaining additional assistance. However, if the team lacks certain skills and abilities to meet project goals, then consider outsourcing. Exchange a person's knowledge and abilities with another team to help one another out.

Be sensitive to time management. Prepare a schedule collaboratively and be realistic in the team's consideration of the time required to meet goals and deadlines. Give positive feedback to one another regularly when someone has completed a task that moves the schedule along. We all appreciate an "atta boy" or "atta girl" every now and then.

Above all, talk and negotiate with one another and try to identify solutions to problems that arise. State issues positively. Instead of describing why something cannot be done, take the upbeat route. Follow the Johnny Mercer lyrics:

> You've got to accentuate the positive
> Eliminate the negative
> And latch on to the affirmative
> Don't mess with Mister In-Between

Discuss what can be done and what each team member is willing to do. Come to a realistic assessment of what can be done within the team's technical, budget, and time considerations. Encourage the expression of differing viewpoints and promote honest dialogue. Express thoughts in a way that does not assign blame.

Put personality conflicts aside and work effectively with one other for the duration of the project. Regardless of negative personal feelings towards a team member, get over it! Team members don't have to love one another but they do have to be cordial and cooperate with one another. Team members should maintain a high level of flexibility and perform a variety of tasks as needed. After the project is over, the team can move on with their life. Remember that the capstone project is a limited duration activity. Getting it done is everyone's priority!

Cultural Styles and Conflict

Team members in industry work in increasingly diverse environments: in terms of age, gender, race, language, sexual preferences, and nationality. Beyond these differences, there are also deep cultural differences that influence the way conflict is handled.

College, university, and workforce teams are increasingly diverse. The team environment may consist of people of different races, religions, nationalities, economic backgrounds, and speaking different native languages. Cultural differences may influence the way solutions are approached. Culture may be defined as the shared set of values, beliefs, norms, customs, attitudes, behaviors, and social structures that guide people's interactions daily.

Some cultures value the group (collectivists) above the individual (individualists). Group conformity and commitment is maintained at the expense of personal interests. Harmony, getting along, and maintaining "face" are thought of as crucial.

The dominant culture in the USA, Canada, Western Europe, Australia, and New Zealand is individualistic, while collectivism predominates in much of the remainder of the world [27].

Individualists and collectivists view conflict differently. Collectivists place a high value on getting along and may view conflict as a sign of social failure. Their society has a low comfort levels with conflict situations—especially of an interpersonal nature.

While many individualists also feel discomfort with conflict, it is regarded as an inevitable part of life that must be dealt with. Conflict with another team member is not necessarily something about which to be ashamed. Understand that it will happen and deal with it. What are the cross-cultural differences in your team? How does it affect your team's performance? How does the team deal with these differences?

Hofstede suggests five dimensions to national culture that may help a cross-cultural team understand one another [28, 29]. These can be summarized as:
1. Hierarchy: Some cultures and consequently possibly a team member emphasize the leader. Individuals may expect the team leader to provide direction and make decisions. Individuals within these cultures tend to be accepting of rules and may not question authority.

At the other end of the continuum are cultures that place a lot of emphasis on team involvement, with wide consultation and group decision-making being common. Questioning authority is likely to be accepted or even encouraged in these cultures.
2. Ambiguity: At one end of the continuum are cultures that encourage risk taking; in these cultures individuals are likely to feel very comfortable trying new and different ways of approaching things. At the other end of the continuum are cultures that place more value on routine, regulation, and formality. Individuals in these cultures are likely to prefer tried and tested ways of doing things rather than taking risks with unknown methodologies.
3. Individualism: This dimension relates to the extent to which the individual values self-determination. In an individualistic culture people will place a lot of value on individual success and the need to look after oneself. At the other end of the dimension are collectivist cultures in which individuals will place more value on group loyalty and serving the interests of the group.
4. Achievement-orientation: Hofstede [28, 29] describes one end of this dimension as masculine and the other end as feminine because it relates to values that have traditionally been associated with gender in western society. A culture at the masculine end of the continuum will be very achievement-oriented, valuing things such as success, achievement, and money. At the other end of the continuum are cultures that place more value on aspects such as quality of life, interpersonal harmony, and sharing.
5. Long-term orientation: At one end of the continuum are cultures that focus on long-term rewards; at the other end are cultures that are more concerned with immediate gain.

A country's cultural values are reflected along a continuum of these five dimensions [28]. Individuals' expectations and behaviors are likely to be influenced by their country's cultural values. For example, according to his research, team members from a country that is high on individualism are likely to:

- Expect to take a role in deciding the team's direction.
- Be prepared to question a team leader's decisions.
- Feel comfortable trying different approaches.
- Focus on achieving their own personal goals, with the view that successful completion of those will facilitate group success.

On the other hand, team members from a country that is low on individualism are likely to:

- Expect a clear hierarchical team structure, with a clear team leadership.
- Be highly disciplined.
- Focus on providing support to other team members to ensure that the overall team outcome is achieved.

The dominant value system of a country is not always at the extreme of one end of the continuum. For example, one country in Hofstede's [28, 29] research falls somewhere in the middle of the power distance and uncertainty avoidance dimensions, suggesting a strong need for hierarchy, but also a tendency to break rules when needed.

Hofstede's [28, 29] framework is useful to help us to think about how misunderstandings may occur within work teams. For example, imagine an international team that is being led by someone from a cultural background that values hierarchy. That leader may expect to make decisions without consulting her/his team members. This may damage the relationship with those team members from cultures at the opposite end of the continuum who expect to be consulted and make joint decisions.

Nonnative speakers of a language may have difficulty working in a second language. Native speakers may understand more definitions for words than second language speakers, which means nuances may be lost. In areas such as engineering, manufacturing, and regulatory compliance, these nuances may be quite important. Regional language, jokes, jargon, acronyms, and sports analogies may be difficult for a nonnative speaker to grasp. Nonnative speakers may require more time to respond to the discussion. If possible hold face-to-face meetings to help gain insight into nonverbal cues that might indicate that a person has a difference of opinion or has some additional thoughts about a subject. Intended or not tone of voice, intonation, body posture, body gestures, facial expressions, and pauses between words conveys information. Successful team integration requires acceptance and understanding of cultural differences within the team while focusing on common objectives.

Conflict will arise from time to time. How the team chooses to respond can be the difference between project success and failure.

Ethics in Project Teams

Project teams require open communication. Participants must recognize their own biases and control them. Honesty in every way must be the byword. Team members should not mislead stakeholders by omission or vagueness. Don't hide behind jargon. If a question is asked to which you don't know the answer respond by saying that you will research it—if you think the point is valid. But don't commit to something that you will not do. The integrity of the team's operation requires each member to step up and take responsibility for his or her work and actions. It will take a long time for the team to regain confidence lost as a result of a team member's misleading or mistrustful behavior. Be guided by the idea of doing the right thing.

Encourage open reporting of "bad news." There cannot be ethical teamwork where individuals are afraid to speak up. Use positive reinforcement. Thank people for notifying you of an error. Each team member should strongly encourage ethical behaviors, such as refusing to allow derogatory remarks in any form. Think about what it means to be a good team player. A set of guidelines might include the following:

- Attend meetings. Show up on time and be prepared to contribute to activities.
- Thoroughly complete tasks and submit work according to the agreed upon schedule.
- Assist other teammates, when asked.
- Listen.
- Participate in team deliberations.
- Respect individual differences (ethnicity, gender, religion, politics, etc.).

- Solve problems in a positive manner.
- Demonstrate reliability.
- Accept responsibility for your actions.
- Communicate transparently—be truthful.
- Accept continuous improvement—be open to new ideas and better ways to do things, consistently examining your methods and welcoming feedback to "find a better way."
- Treat others with dignity, fairness, respect, and courtesy.
- Know when to stop advocating for your position. It may hurt to have your pet idea rejected by the team after all that honest effort. However, don't let hurt feelings goad you into talking badly about teammates. Conversely, don't think or talk negatively about yourself if the team chooses a different idea or direction.
- Honor and maintain the confidentiality and privacy of colleague, customer, client, and employer information.

Project Team Peer and Self Evaluation

A positive attribute of the capstone team process is that teamwork competencies can be acquired during team-based activities. Classes with lectures focusing on individualized tasks do not afford this benefit. Who knows the team better than its members? Who better should evaluate the team than the members of the team? Grading a team is difficult at best. If only the faculty mentor evaluates the team it may not be a fair reflection of an individual's work. After all, a significant portion of the work is completed outside the classroom. Consequently, some faculty have begun to use a peer review and self-evaluation instrument as a major component of the grading rubric. The CATME project began in 2003 with the development of an instrument for self and peer evaluation and team management called the Comprehensive Assessment of Team Member Effectiveness (https://engineering.purdue.edu/CATME/). Others have also sought an effective evaluation instrument to assess individual contributions and overall team performance (see for example [30–32]).

The faculty mentor should inform team members at the beginning of the semester that a peer evaluation will take place that will emphasize team process and individual contributions in addition to project performance. These measures are different from most courses that student's take. Keep in mind that evaluation is a qualitative judgment intended to provide feedback for improvement. The evaluation instrument is a tool to foster teamwork competencies such as communication, leadership, collaboration, and interpersonal relations. These qualities can be acquired during team-based activities. A mid-semester process check as well as an end of semester evaluation can be used to assist the team in improving its performance.

The Peer and Team Evaluation Instrument in Table 2.1 is adapted from material developed by Wilson [30]. The evaluation criteria in Table 2.1 address individual contributions to the team. Before using the instrument, the class should discuss and modify, add or delete criteria based on class consensus.

Table 2.1 Peer and team evaluation instrument

No.	Task	Self	Student name 1	Student name 2	Student name 3
1	Attendance at team meetings: Present at all team meetings except where a previous commitment conflicted with the time and the absence was agreed upon with team members				
2	Planning and Task definition: Helped develop and support the Work Breakdown Structure (WBS) and schedule				
3	Technical support: Provided technically creative and insightful ideas to the effort				
4	Timeliness: Completed all assigned tasks in a timely manner				
5	Collaboration: Made a genuine effort to work effectively with others. Shared ideas openly with fellow team members. Open-minded, objective, respected other's ideas, positive				
6	Effort: Exhibited a high level of interest and commitment to the project				
7	Contribution of skills: Obtained results using competencies, resources, and materials				
8	Contribution of ideas: Provided creative and innovative ideas for group discussion				
9	Oral Communication: Spoke clearly, succinctly during presentations and responded knowledgeably to questions				
10	Written Communication Contributions: Preparation of the specification				
11	Written Communication Contributions: Preparation of presentations				
12	Written Communication Contributions: Preparation of the preliminary proposal				
13	Written Communication Contributions: Preparation of the final project documents				
14	Test: Contributed to the product, service, process, or system test				
15	Problem solving: Defined issue. Set priorities. Developed and implemented solution. Monitored progress and adjust direction as needed				
	Total				

Rating Scale:
1. Did not contribute in this way.
2. Willing, but contribution not very successful or useful.
3. Average contribution. Did what was required.
4. Significant or above average contribution. Did more than required.
5. Outstanding contribution. Did much more than was required. Made a great difference to the team and to the project.

Other Comments:

Each peer and self-evaluation criteria was assigned a five-point Likert scale from almost never (1) to almost always (5). Team members should insert a value (from 1 to 5) into the appropriate box for each of the 15 items measured. The maximum individual score is 75. Write in each team member's name across the top including your own, and then rate each person in the categories using the number system given. This evaluation must be done anonymously and in private. The team should not see the ratings until all team members have completed the evaluation. Be honest with yourself and in evaluating others. Very few people are all 1's or all 5's. The results of the peer review can be given to the faculty mentor who would then distribute the information to each team member to be read in private. Alternatively, the faculty mentor can post the evaluation instrument on the Web site SurveyMonkey.com and request each team member to complete the evaluation form. The anonymous responses can then be downloaded by the faculty mentor and distributed to the team for their review.

When completing the evaluation, disregard your general impressions and concentrate on one category at a time. Carefully review the category description. Think about instances that are typical of each team member's work and behavior. Do not be influenced by unusual situations that may not be typical. Determine the rating that best describes the team member's accomplishments in that area and enter the selected rating number. If a factor has not been observed during the rating period, enter NA for not applicable. If team members wish, comment at the end of the evaluation to further describe a rating.

Self- and peer-evaluation pushes students to take responsibility for their effort and participation and, therefore, for the success of their team. Peer review leads to collegial feedback and reflective thought and action. Team members are in the best position to comment on peer effort, quality of technical content, materials, class presentation. Evaluations which are conducted during the capstone learning experience are often called formative. Evaluations which are conducted at or near the end of the capstone learning experience, and which provide a retrospective view of the overall value of that experience are summative. In all cases, the evaluation results indicate the changes, if any, that members of the team need to make.

The peer review process is difficult. We all take pride in the work we do and the products we create. We don't like to admit that we are not as good as we think we are and we don't like to have other people tell us about our shortcomings. Conducting successful self and peer reviews requires us to overcome this natural resistance to critique others and ourselves. We all have egos and team members must demonstrate compassion and sensitivity for colleagues during the review process. Team success depends on helping each other do the best job possible.

Basic Team Communications   3

> *Engineering is not merely knowing and being knowledgeable, like a walking encyclopedia; engineering is not merely analysis; engineering is not merely the possession of the capacity to get elegant solutions to non-existent engineering problems; engineering is practicing the art of the organized forcing of technological change… Engineers operate at the interface between science and society…*
> —Dean Gordon Brown; Massachusetts Institute of Technology

> *Engineering is the science and art of efficient dealing with materials and forces…it involves the most economic design and execution…assuring, when properly performed, the most advantageous combination of accuracy, safety, durability, speed, simplicity, efficiency, and economy possible for the conditions of design and service.*
> —J. A. L. Waddell, Frank W. Skinner, and H. E. Wessman

Chapter Objectives

After studying this chapter, you should be able to:
Understand a meeting's purpose.
Prepare a meeting agenda.
Manage the meeting follow-on process.
Participate in a brainstorming session.
Prepare and deliver a quality presentation.

Conducting meetings

Like it or not the team will have many meetings during the capstone course. Every team member has participated in meetings in school, in the family, in a place of worship and in business activities. Most meetings fall into one of the following categories:
1. To gather or provide information (e.g., receive reports, find facts).
2. To exchange ideas, views, opinions, and suggestions.
3. To discuss options.
4. To find solutions to problems.
5. To make decisions.
6. To devise plans.
7. To initiate an action.
8. To divide responsibilities.
9. To follow up on actions taken and their results.
10. To inspire and organize individuals for and against issues.
11. To promote products or events.
12. To protest against issues or actions.

Try to keep the number of topics to be considered to no more than two or three at each meeting. It is difficult to focus on too many issues at any one time.

At the beginning of the project, people meet to determine one another's interests and competencies. Soon after that the team will identify and discuss possible topics for its work. Select a time and place for a regularly scheduled gathering. An online meeting is fine as well but at the beginning hold face-to-face meetings to facilitate the team getting to know one another. Meet for the purpose of accomplishing tasks that will move the team forward. If the team meets out of a sense of obligation or habit—so be it. It is important that team members get in the habit of meeting at least once a week. Regularly scheduled meetings will motivate team members to complete an assigned task—if only to avoid embarrassment in front of the team. Procrastination is a deadly enemy for timely project completion. Enthusiasm, sincerity, and good intentions are wonderful but only results that move assigned tasks along are important.

Meeting Preparation

Prepare an agenda for each meeting and distribute it to the expected participants before the meeting. Knowing the information to be discussed permits the attendees to prepare and bring supporting material to the meeting. Solicit agenda items from the team in advance. Consider a rotating meeting leadership. It may be helpful to have each team member take a turn facilitating and managing the meeting—that is, setting the agenda, setting the ground rules, introducing topics, selecting speakers, and managing time. The facilitator assigns a person to take minutes. A typical project meeting agenda is shown in Table 3.1.

Table 3.1 Project meeting agenda

Meeting Agenda

Date:
Meeting purpose:
Start time:
Expected end time:
Location:
Items to bring with you:
Invitees:

Topics

Announcements
Review minutes of previous meeting
Project Status Review – Each team member participates by reporting on the following:
    Review actions items assigned at the last meeting
    Accomplishments/activities since last review
    Plans for next period
    Issues resolved
    New issues encountered and if necessary, mitigating plans
    Other comments or concerns
Review-assigned action Items

Note that the agenda includes a start time and an end time. People's time is valuable and must be respected. Meetings should begin on time and end on time. If the participants are not all there at the start time, then begin without them. A small project team demands cooperation by all and barring significant unforeseen circumstances, team members should arrive on time and leave at the meeting's end. The agenda should be realistic and the meeting facilitator has to be ruthless in respecting people's time and making certain that the meeting ends as planned. Table issues that cannot be resolved, but make certain that a person is charged with obtaining additional information so that the issue can be discussed at another meeting. If time is running out in the midst of a fruitful discussion, ask the team members to vote on whether or not to go beyond the scheduled end time.

The facilitator should guide the discussion and discourage disruptive or negative behavior. At the end of the meeting summarize the action items and identify those responsible to complete the items and the due dates.

After the Meeting

Prepare short meeting minutes as a reminder to all of the items discussed, actions agreed upon, and assignments due including what must be done, who has responsibility, and when it is due. Circulate the minutes as soon as possible after the meeting has taken place. A typical meeting minutes template is shown in Table 3.2.

Table 3.2 Project meeting minutes template

Meeting Minutes			
Date:			
Location:			
List of Attendees:			
Key issues or discussion:			
Discussion item 1			
Discussion item 2			
Discussion item 3			
Action Items:			
Description of Action Item	Person Responsible	Date Due	Submit Results To
Next Meeting Date:	Time:	Location:	

Thinking About the Project Topic

The team will spend anywhere from 4 to 8 months working together on a project of mutual interest. Ideally the topic should be a natural outgrowth of team members' interests and combined skills. The team members will use their cumulative knowledge from formal classes, work experiences and hobbies. Teams typically choose a topic involving a product, service or process to implement. Many institutions encourage students' to select a topic involving a community-based service. Naturally, avoid choosing topics that might endanger team members or others safety.

As college and university resources are limited, the team will be expected to make efficient use of time, money, people, equipment, and facilities. Teams should avoid choosing topics that might involve extraordinary expenses. Discuss funding with the project's faculty mentor. Be flexible as the team will have to make contingency plans for unexpected circumstances. Change will happen.

Brainstorming

Sometimes a team is formed based on a common interest in a topic. Other times, the faculty mentor assigns teams or friends get together and form a team without having a clearly defined area of concentration. In the latter case, they must select a topic of common interest. After they identify an area of interest they identify prospective solutions. The selection of a topic and the identification of a proposed solution require a free information exchange and flow of ideas. To do this, teams frequently engage in a brainstorming session. Brainstorming is a process of open-ended idea generation. Brainstorming is a process for developing creative solutions to problems. Alex Osborn, an advertising manager, is credited with developing the method [33]. Research shows that brainstorming tends to improve team performance [34].

Brainstorming works by focusing on a problem, and then coming up with as many solutions as possible and by pushing the ideas as far as possible. Participants not only come up with new ideas, but also develop and refine one another's ideas. Traditional face-to-face brainstorming centers on individuals interacting. Electronic brainstormers can present ideas simultaneously. In electronic brainstorming, group members generate their ideas and then look at the group contributions when they exhaust their own ideas [35].

Whether face-to-face or electronic, Osborn [36] suggests that in order to be maximally productive, brainstorming groups should follow these rules:

No criticism: Do not criticize one another's ideas. This permits team members to feel comfortable with the idea of generating unusual ideas without disparagement.

Come up with as many ideas as possible: Out of the box, wild ideas are OK. It is easy to tone ideas down. The greater the number of ideas, the more the likelihood of winners.

Based on Osborn's work, the following steps have evolved and have been used by many teams to conduct a successful brainstorming session.

1. Gather the participants from as wide a range of disciplines with as broad a range of experience as possible. This brings many more potentially creative ideas to the session.
2. Write down a brief description of the problem—the leader should take control of the session, initially defining the problem to be solved with any criteria that must be met, and then keeping the session on course.
3. Use the description to get everyone's mind to zoom in on the problem. Post the description. This helps in keeping the group focused.
4. Encourage an enthusiastic, uncritical attitude among brainstormers and encourage participation by all members of the team. Have fun!
5. Write down all the solutions that come to mind. Do NOT interpret the idea; however team members may rework the wording for clarity's sake.
6. Do NOT evaluate ideas until the session moves to the evaluation phase. Once the brainstorming session has been completed, the results of the session can be analyzed and the best solutions can be explored either using further brainstorming or more conventional methods.
7. Do NOT censor any solution, no matter how silly it sounds. The silly ones will often lead to creative ones—the idea is to open up as many possibilities as possible, and break down preconceptions about the limits of the problem.
8. The leader should keep the brainstorming on subject, and try to steer it towards the development of some practical solutions.
9. Once all the solutions have been written down, evaluate the list to determine the best action.

After the brainstorming session, the team needs a more serious meeting to evaluate and qualify the ideas generated. Devise standards for evaluating and revising the collected ideas. Then see how many ideas fit the criteria. Discuss methods or ways to develop and prioritize the ideas. Some ideas might need to be set aside for future consideration. This phase is the beginning of where raw ideas will be forged into product, service, or process development initiatives.

As discussed in Chap. 3, a team in the Tuckman Norming stage offers great benefits for getting work done. However, the team may inadvertently prevent the brainstorming process from taking place. A very closely knit team may enter the groupthink "mode" which would stifle individual creativity and independent thinking in group members. Remember, groupthink occurs when a team makes faulty or ineffective decisions for the sake of reaching a consensus. Presenting and debating alternative viewpoints are sacrificed for team cohesion, which can significantly hinder the decision-making and problem-solving abilities team. Be sensitive to this. The team does not want to disregard realistic alternatives in an effort to maintain harmony. The brainstorming process is there to increase chances for successful idea generation.

Presentations

> According to most studies, people's number one fear is public speaking. Number two is death. Death is number two. Does that sound right? This means to the average person, if you go to a funeral, you're better off in the casket than doing the eulogy."—Jerry Seinfeld

Preparing for the Presentation

Nervousness or anxiety associated with public speaking is common. Of all social situations, public speaking is most feared by the general population [37, 38]. So while fear of public speaking is a common phobia, it must be overcome. Men and women in business situations must speak regularly with colleagues, customers, vendors, at status meetings and at board meetings. But with preparation, courage, and support from peers you can overcome your fear. During the capstone course team members will get the opportunity to speak weekly, so everyone will get to practice.

Consider the following ideas to help in presentations.

- Prepare and know the topic. The better you understand what you're talking about, the less likely you will make a mistake or get off track. Think about the questions your faculty mentor or audience may ask and have responses ready.
- Practice the complete presentation several times and then practice some more. For a presentation that will be given before a large group, record it with a video camera and watch yourself so you can see opportunities for improvement. When making your presentation listen to yourself. Stop yourself before you say "umm," "uh," "like," and "you know." These words will detract from your speech. Watch the video to see how you did.
- Take two or more deep, slow breaths before beginning the presentation.
- Focus on the material, not on the audience. People are primarily paying attention to the information you are presenting—not how you are getting it across. Chances are they will not notice minor mistakes. If audience members do notice you are nervous or that you get a little off track, they won't judge you. They are rooting for you and want your presentation to be a success.

- Pause for a moment if you lose track of what you are saying or your mind goes blank. Your audience will likely not mind a pause to consider what you have been saying.
- Dress in a way that will enhance the message of your presentation. In business and industry the unwritten guide is to dress for the position that you want [39]. Dressing in shorts or jeans with holes or knee patches is not the ideal way to command respect—even in a classroom.
- The audience will detect a speaker's confidence through their posture, voice, facial expressions, gestures, and how they interact (or don't interact) with them.
- Do not read the PowerPoint slides word for word. Never read a technical paper word for word. The audience can read as well as you. They seek further understanding of the contents from the speaker.
- Be careful of the use of jargon but speak the audience's language. Define unusual words. Identify the meaning of acronyms.
- Face the audience and not the screen. Of course, do not stand in front of the screen.
- Make eye contact with one person at a time.
- Speak loudly and distinctly! Do not mumble.
- Vary the vocal tone and pitch enough to keep listeners interested.
- Direct your words to all parts of the room.
- Avoid ending sentences in an escalating tone that suggests every sentence is a question.
- Discreetly pay attention to the allotted time that you were given.
- Make sure that all members of the team have a role.

Powerpoint Presentations

Keep your PowerPoint slides simple. Don't overload them with details. The slides are not a place to dump as much data as possible. Instead, they are a visual communication aid. The emphasis should be to deliver a clear and meaningful message. Most engineers seem to relate to visual representations. So use graphics to illustrate points. Diagrams can frequently highlight information better than text. *Excel or Word charts can be copied and pasted on a PowerPoint slide. But make sure the graphs and charts are legible.* If helpful, use a free stock image which may be available from websites such as openclipart.org, Everystockphoto.com, clker.com, stockphotosforfree.com, or pixabay.com.

A good PowerPoint presentation includes:
- An initial slide with the presentation title, date of presentation, speaker's name, and contact information.
- Numbered slides.
- The name of the conference or company that the information was delivered to in a conspicuous place.
- An agenda that tells the audience where you intend to go. Use topic headers at the top of your slides. Include a summary conclusions slide. The more you help an audience know where you are going, the more they will stay with you.

- Charts, graphics, photos, and other images to make a point. But remember that art work and animation may distract your audience. Art work does not substitute for content. Don't try to dazzle the audience with animations, graphics, or style… but with the information. Sound effects may distract too. Use sound only when necessary.
- Graphs rather than just charts and words. Data in graphs are easier to comprehend and retain than are raw data. Trends are easier to visualize in graph form. Always title your graphs and label axes (include units).
- Word colors and underlines used sparingly and only to emphasize points.
- Bullets or numbers to show a list or order.
- Layout continuity from slide to slide.
- Headings, subheadings, and logos that show up in the same spot on each slide.
- Margins, fonts, font size, and colors located in the same general position on each slide.
- Consistent use of lines, boxes, borders, and open space throughout the presentation.
- Left justification of most bullet points to keep the slide neat and easy to follow. Centering several lines of text makes the text ragged and may be hard for the audience to read.
- Contrasting colors. For example, use a dark background and white letters. Select a design template where the words are easily distinguished from the background. Keep the background simple.
- Font sizes equal to or greater than 20 points with no more than ten sentences per slide.
- Different size fonts for main points and secondary points.
    - For example use a main point font of 28-points, a secondary point font of 24 points and a title font of 36-points
- A standard font like Times New Roman or Arial. Do not use a complicated font. If you insist on being different you may want to go to Smashing Magazine (http://www.smashingmagazine.com/2012/07/20/free-font-round-up/), which offers free fonts.
- Upper case (capital) letters for words only when necessary. Always use capital letters for acronyms. Continuous use of upper-case letters makes every word a rectangle, which takes much longer to read and becomes uncomfortable and tiring.
- Italics for "quotes," or to highlight thoughts or ideas, or for book, journal, or magazine titles.
- A minimum of transition effects between slides.
- Source citation.
- A strong closing slide. The audience is likely to remember your last words. Use a conclusion slide to:
    - Summarize the main points of your presentation
    - Suggest future avenues of research

Technical and business writing uses acronyms all the time. Acronyms are shorter forms of phrases that are useful when repeating the same words several times. For example, IEEE represents the Institute of Electrical and Electronic Engineers or OSHA stands for the Occupational Safety and Health Administration. The first time an acronym is used, the words should be written out with the short form placed in parentheses immediately after. For example, write the phrase Occupational Safety and Health Administration (OSHA). The audience will then be aware that any future reference to OSHA in your writing refers to the Occupational Safety and Health Administration. After establishing an acronym, consistently use that acronym in place of the words. Writing the acronym saves time and space. If many acronyms are needed, consider including a glossary of acronyms placed at the end of the presentation.

Take care to present the argument methodically. Include an overview or agenda followed by the body and always present a conclusion. Double check the presentation for spelling and grammatical errors. If English is not your first language, have someone else check the presentation for sentence structure, spelling, and grammar!

Number your slides so that a member of the audience can call out a slide number if they have a question. As a guide, plan on 1–2 slides per minute of your presentation. Write in a bullet format, not complete sentences. Avoid wordiness, use key words and phrases only. Use a wireless slide advance device and a laser pointer. This enables the speaker to move about the room.

It is important to be transparent about your sources. Provide citations for sources used within the presentation. There are two common methods used to create citations for use within PowerPoint. Either provide the source on the slide that uses the citation or provide a reference page at the end of the presentation.

Figure 3.1 provides a guide to presenting information in a good light. Poorly designed presentations can leave an audience feeling confused and bored. Review these Do's and Don'ts for ideas on making your presentation clear.

Practice your presentation. Never read your presentation word for word. The audience came to see and hear you. Use the slides as a structural guide. The media should enhance the presentation, not be the presentation. If you only read from the slides, then you could just as well have sent the audience copies of the slides! Only you can prevent "Death by PowerPoint."

Speak slowly, loudly, and clearly. Make sure that the ends of your sentences don't drop off or rise as if asking a question. Maintain eye contact. It conveys confidence and openness. It also lets you know how the audience is responding to your presentation. In a large audience make eye contact with different people in different sections of the room on a rotational basis.

Use hand gestures naturally and to emphasize points. When not gesturing, let your hands drop to your sides naturally. Keep them out of pockets, off your hips, or behind your back. Avoid playing with clothes, hair, or presentation materials. Maintain good posture, but do not be rigid. Occasionally move from one spot to another, stop, and then continue to speak. Don't pace.

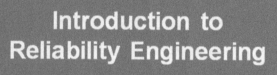

a - A clear presentation title page states:

- Presentation subject
- Author or team member names
- Contact information
- Class name and number
- Date of the presentation
- Institution's name

The Arial font is simple and easy to read.

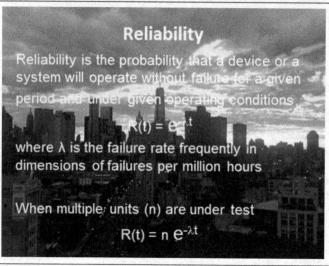

b - Although a beautiful picture of One World Trade Center (the Freedom Tower) as seen from lower Manhattan at dusk, the cluttered background detracts from the slide contents.

Fig. 3.1 Presentation do's and don'ts

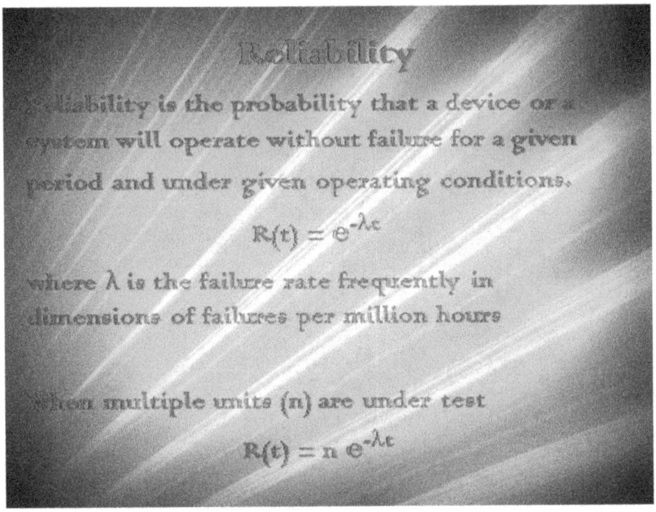

Presentations

Slide (top):

Reliability

Reliability is the probability that a device or a system will operate without failure for a given period and under given operating conditions.

$$R(t) = e^{-\lambda t}$$

where λ is the failure rate frequently in dimensions of failures per million hours

When multiple units (n) are under test

$$R(t) = n\, e^{-\lambda t}$$

c - The slide represents a poor choice of both background and font style. In addition, it may be particularly difficulty to see for those people having red-green color blindness.

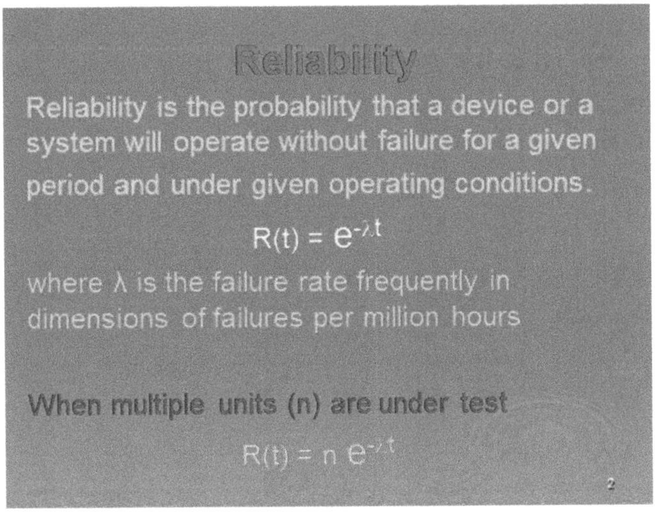

Slide (bottom):

Reliability

Reliability is the probability that a device or a system will operate without failure for a given period and under given operating conditions.

$$R(t) = e^{-\lambda t}$$

where λ is the failure rate frequently in dimensions of failures per million hours

When multiple units (n) are under test

$$R(t) = n\, e^{-\lambda t}$$

d - Multicolored print may look pretty but it distracts from the information content. It also detracts from the importance of the presentation.

Fig. 3.1 (continued)

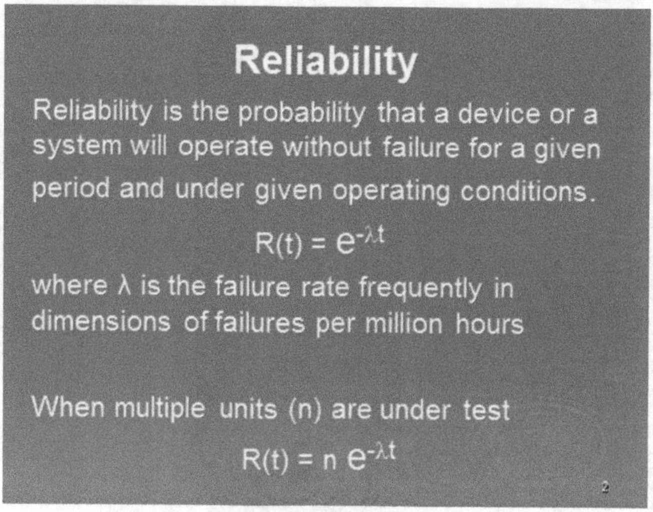

Reliability

Reliability is the probability that a device or a system will operate without failure for a given period and under given operating conditions.

$$R(t) = e^{-\lambda t}$$

where λ is the failure rate frequently in dimensions of failures per million hours

When multiple units (n) are under test

$$R(t) = n\, e^{-\lambda t}$$

e - A plain background with contrasting print permits the slide contents to be easily seen. Eight information lines using 32 point font size enhances readability. The slide number is shown on the bottom right.

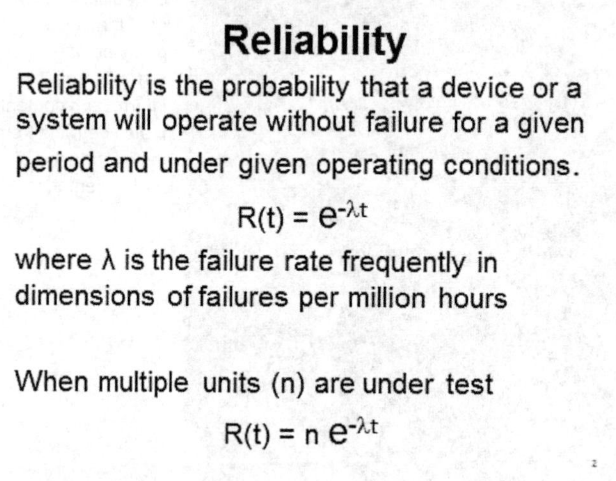

Reliability

Reliability is the probability that a device or a system will operate without failure for a given period and under given operating conditions.

$$R(t) = e^{-\lambda t}$$

where λ is the failure rate frequently in dimensions of failures per million hours

When multiple units (n) are under test

$$R(t) = n\, e^{-\lambda t}$$

f - Good contrast with the background makes text easy to read.

Dark text on a light background is fine. Dark backgrounds are effective, if the text is light.

Patterned or textured backgrounds may make text hard to read.

Keep the background and font color scheme consistent throughout the presentation.

Fig. 3.1 (continued)

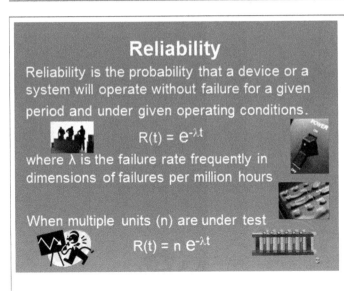

g - Use photos, clip art, charts and diagrams to emphasize presentation key points. When used incorrectly, as in this slide, excessive artwork only confuses the audience and detracts from the presentation. Ensure that the images are relevant to the slide's content; otherwise, they only distract the audience from the important content.

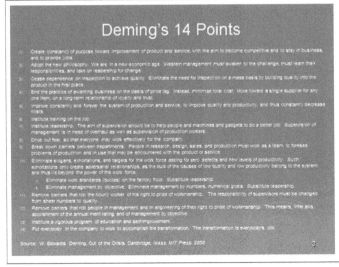

h - The slide is numbered and includes a source for the material. However, the font size is so small that the material is unreadable. It should be separated into two or three slides.

Fig. 3.1 (continued)

Intel Microprocessor History

Microprocessor Device	Year Introduced	No. of Transistors
4004	1971	2,250
8008	1972	2,500
8080	1974	5,000
8086	1978	29,000
80286	1982	120,000
80386	1985	275,000
80486	1989	1,180,000
Pentium® processor	1993	3,100,000
Pentium® II processor	1997	7,500,000
Pentium® III processor	1999	24,000,000
Pentium® 4 processor	2000	42,000,000
Pentium® M processor	2003	55,000,000
Core™2 Duo processor	2006	291,000,000
2nd generation Core™ processor	2010	1.16 billion
3rd generation Core™ processor	2012	1.4 billion

Source: http://www.intel.com/content/www/us/en/history/museum-transistors-to-transformations-brochure.html

i - The font (Arial) sizes used in this slide are:

Title – 36 points

Body – 20 points

Source – 12 points

The chart is very crowded and the material may not be able to be seen by all in the audience.

Intel Microprocessor History (1)

Microprocessor Device	Year Introduced	No. of Transistors
4004	1971	2,250
8008	1972	2,500
8080	1974	5,000
8086	1978	29,000
80286	1982	120,000
80386	1985	275,000
80486	1989	1,180,000

Source: http://www.intel.com/content/www/us/en/history/museum-transistors-to-transformations-brochure.html

j - The font (Arial) sizes used in this slide are:

Title – 36 points

Body – 28 points

Source – 12 points

The "eye chart" has been eliminated. The previous slide is divided in two and the information can be easily seen.

Intel Microprocessor History (2)

Microprocessor Device	Year Introduced	No. of Transistors
Pentium® processor	1993	3,100,000
Pentium® II processor	1997	7,500,000
Pentium® III processor	1999	24,000,000
Pentium® 4 processor	2000	42,000,000
Pentium® M processor	2003	55,000,000
Core™2 Duo processor	2006	291,000,000
2nd generation Core™ processor	2010	1.16 billion
3rd generation Core™ processor	2012	1.4 billion

Source: http://www.intel.com/content/www/us/en/history/museum-transistors-to-transformations-brochure.html

k – The second half of the chart shown in Figure i). The reference website need not be in a large font size. After receiving a copy of the slides, the listener may check the source of the information after the discussion has completed.

Fig. 3.1 (continued)

Capstone Class Written and Oral Submittals

> *Life is like riding a bicycle. To keep your balance, you must keep moving.*
> —Albert Einstein
>
> *Engineering is the conscious application of science to the problems of economic production.*
> —H.P. Gillette

Chapter Objectives

After studying this chapter, you should be able to:
Identify the capstone course phases.
Avoid common report writing errors.
Gain insight into the project selection process.
Use a SWOT tool to evaluate a project.
Understand the required written and oral deliverables expected in the capstone course.
Recognize the importance of market research in project selection.
Prepare a list of project tasks.
Understand the importance of finances in a project.
Prepare a project budget.
Use a weekly status log.
Prepare a weekly project status report.
Evaluate project risk.
Prepare and use a project specification.
Prepare a project proposal and final report.

As described in the preface, the capstone course may be divided into five phases, which are
- Team selection
- Project selection
- Proposal preparation and presentation
- Project development
- Final report, presentation, and demonstration

Each of these phases has a set of objectives and deliverables that comprise the project process. A common thread throughout the phases is writing—whether for PowerPoint slide presentations or reports. Most students reading this book are engineers and might think that writing is unimportant in their career. Nothing could be further from the truth. Writing well is important for success in any business or organization. Report writing is a primary professional responsibility of the practicing engineer. People who can write clearly and concisely have a competitive edge over others who struggle to communicate.

Business or technical writing differs from literary writing. The main purpose of business and technical writing is to inform—not entertain. The style uses simple, specific, and precise language. Never use a long word where a short one will do. The writer should try to transmit information as objectively as possible.

When composing an e-mail, include the major thought in the first paragraph. Try to use active verbs instead of passive verbs. (Passive verbs are forms of to be.) Beware of common grammatical mistakes, like subject–verb agreement. Singular subjects need singular verbs; plural subjects need a plural verb. Use a singular verb form after nobody, someone, everybody, neither, everyone, each, and either. Know when to use "that" and "which." "That" introduces essential information to the meaning of the sentence. "Which" introduces extra information. Another common error is confusing "affect" and "effect." Affect is a verb meaning "to influence." "Effect" is a noun that means "result." Here are some other common potentially ambiguous word combinations [40]:

Virgule
The virgule is a diagonal mark (/) used to separate alternatives, as in and/or or miles/h. However, the use of the virgule can be confusing. It may leave the reader free to choose whether the sentence ought to read "and" or to read "or," whichever reading is cheapest or easiest to satisfy.

As a minimum" and "not limited to"
Clearly spell out what is meant.

Ensure, insure, and assure
Ensure means to make certain.
Use insure only when you mean "to issue or buy an insurance policy."
Assure means to state something sincerely.

Ambiguous adjectives and adverbs

Words that allow a broad range of interpretation may get you into trouble, because they may be interpreted however the reader sees fit. Listed below are some vague qualitative adjectives and adverbs. Use them carefully in your writing.

ABOUT	ACCEPTABLE	ACCURATE	ADEQUATE
ADJUSTABLE	AFFORDABLE	APPLICABLE	APPROPRIATE
AVERAGE	BETTER	CAREFUL	DEEP
DEPENDABLE	DESIRABLE	EASY	ECONOMICAL
EFFICIENT	ESSENTIAL	EXCESSIVE	GOOD
HIGH	QUALITY	IMMEDIATELY	IMPROPER
INSTANT	INSUFFICIENT	KNOWN	LESS
LOW	MAJOR	NEAT	NECESSARY
NORMAL	OPTIMUM	OTHER	PERIODICALLY
PLEASING	POSSIBLE	PRACTICABLE	PRACTICAL
PROPER	QUICK	REASONABLE	RECOGNIZED
RELEVANT	REPUTABLE	SAFE	SECURE
SIGNIFICANT	SIMILAR	SIMPLE	SMOOTH
STABLE	SUBSTANTIAL	SUFFICIENT	SUITABLE
TEMPORARY	TIMELY	TYPICAL	VARIABLE
VARIOUS	WIDE	WORSE	

Good writing skills allow you to communicate your message with clarity to an audience. Checking for correct grammar, word usage, and spelling mistakes is a courtesy to your readers since it can take them much longer to understand the message content if they have to think and reread the text to decipher the correct information. While misunderstanding resulting from poor writing may lead to a lower grade in school, misinterpretation in business may lead to financial loss or worse. This chapter deals with proposal and presentation writing—an area in which engineers are commonly involved.

While on the subject of word clarification, the phrases proof-of-concept (POC) and proof-of-principle (POP) are sometimes incorrectly used interchangeably with prototype or pilot. A proof-of-concept or proof-of-principle is a realization of a method or idea to demonstrate its feasibility. It is a demonstration to verify that a concept has the potential of being used. The demonstration may not meet the form, fit, and function of the prototype or the final product. In engineering and technology, a breadboard of a new idea may be constructed as a POP. The POP might also be used as a means to demonstrate the functionality of an idea before the patent application is filed.

A prototype is a more fleshed out realization of the intended product. The prototype may bear resemblance to the final product. It generally does come close to meeting the form, fit, and function of the final production unit and is used to validate the idea. An operational prototype may also be used to show prospective investors the idea for the purpose of raising money.

A pilot is a limited run of the product using the full production system and tests its usage with a subset of the intended audience. The reason for doing a pilot is to get a better understanding of how the product will be used in the field and to refine it.

Project Selection

Topic Selection: Search Phase

Identifying a project topic is difficult. Choose a subject about which you feel passionate. Select a topic that is narrowly focused and carefully defined. Team members must have a willingness to demonstrate independent skills and become competent in new disciplines. View the capstone course as an opportunity rather than a burden. How can this project act as a stepping stone to a job? Consider a topic that you have previously only thought about but never spent the time to investigate. Do you want to consider a project that will look good on your résumé? Can you find a topic that makes use of not only your academic interests, but also your work experience?

Faculty recognize that identifying a topic for a capstone project is very challenging. It will almost certainly require multiple iterations. Sometimes it helps to formulate a topic as a question to be answered—as that will give the team focus and direction. Regardless of the strategy one uses in identifying a research project, the student should expect to work closely with their faculty mentor; the enthusiastic and willing participation of the advisor is essential to the success of the project.

Start with a need. Then extrapolate from your past experience. Apply fresh thinking to old problems. Development ideas can start in many ways, including:

- Copycat: Sometimes, first-out-of-the-gate products receive competition by later competitors who refine and improve an idea or make it cheaper. However, be careful. A federal grand jury in 2012 found Samsung guilty for, among other things, making phones and tablets whose features and rounded corners copied too closely those of Apple's. Samsung was told to pay Apple damages in excess of $1 billion. In the restaurant business, a chef legally can copy a competitor's strawberry shortcake, so long as the chef doesn't steal the recipe.
- Piggyback: Building onto an existing technology. One strategy to deal with the problem of technological obsolescence is "piggybacking." Piggybacking is a strategy that enables renewed functionality of a technologically obsolete product through the integration or add-on of a secondary device or component [41]. A common piggyback action that people are guilty of is to connect to someone's Wi-Fi network without his or her permission.
- Combine Ideas: Take two ideas and put them together to make one new idea. Find the peanut butter to go with the jelly. The low-tech Snuggie is the mutation of a blanket and a robe into a blanket with sleeves.
- Leapfrog: Some developing countries that lack expensive infrastructures like telephone lines and power grids are bypassing these massive investments and "leapfrogging" directly to lower-cost, decentralized technologies such as cell-phones instead of landlines, satellite TV instead of local transmissions, or solar panels instead of grid electricity. They use solar water heaters, and lighting and traffic signals powered by solar technology.

- Breakthrough: A creative advance that gives people powerful new ways to use technology that are significantly different from prior "art". Think the transistor or the personal computer or 3-D printing or hydraulic fracturing ("fracking").

For the purposes of the capstone course think simple innovative ideas that your team can implement. Don't worry that it may not revolutionize an industry. Most students focus on advanced technology products and services. Sustainable, low-technology solutions to problems for rural areas and developing countries can dramatically increase productivity and improve the quality of life. Examples include human-powered irrigation pumps; recumbent pedal powered flat-bed trucks used as a high-bulk load carrier; the use of a windmill to again convert kinetic energy directly into mechanical energy for farming purposes; and adjustable focused eyeglasses using a fluid-filled lens. Think about the social benefit resulting from working on a project of this nature.

If the team considers a product or service that may be used by a company, it should offer a competitive advantage. The idea should add value to existing initiatives. It should show a significant return on investment to justify the investment. For an industrial organization, an idea must lead to an effective and worthwhile solution or an opportunity.

Take a look around to get ideas for projects. Think of the most frustrating or simply annoying problems you can imagine. Now think about what approach or device or service you could provide to fix them. Are you involved in a club or local organization that has a need for a technical product or service? If you are working, does your company have a need for a product, service, or process change that will automate a task or make something more efficient? Do you have disabled friends or colleagues that could benefit from a technology innovation?

There is a need for improved health informatics systems in hospitals and healthcare agencies that will improve health care delivery in resource-constrained environments. Consider redesigning a standard everyday product to make it more user-friendly. Take into consideration functionality, material required, manufacturing processes involved, and cost. How would you make a process, service, or product cheaper, faster, smaller, bigger, or more reliable?

Perhaps team members would like to address environmental problems and perform a social good. Developing nations and certain areas in the United States require simple and sustainable water purifiers and irrigation systems. Useful environmental projects may involve waste management or the use of wind, solar, or hydro energy sources to power equipment or reduce energy costs. Energy harvesting, that is energy derived from external sources (e.g., solar power, thermal energy, wind energy, salinity gradients, and kinetic energy), captured, and stored for small, wireless autonomous devices, like those used in wearable electronics and wireless sensor networks may represent an area of investigation. Green buildings using indigenous materials that are easy to construct locally are important structural projects. Do you have an idea for an inexpensive desalination technique to convert sea water to drinking water on ships or arid regions of the world?

In parts of the world, live mines remain buried in fields. Developing a device that would detect land mines or naval mines in an area would aid in their removal, which would save lives.

Consider applying Six Sigma or lean methods to a manufacturing process or a health care facility to improve the quality of process outputs by identifying and removing the causes of defects (errors) and minimizing variability in manufacturing and business processes.

The healthcare industry is a ripe area for improving products and processes. The necessity for quality and safety improvement initiatives permeates health care [42]. According to the Institute of Medicine (IOM) report, To Err Is Human [43] the majority of medical errors result from faulty systems and processes, not individuals. A simple example of possible improvements that the engineering world could bring to the healthcare industry deals with fittings called Luer connectors. Luer connectors link various system components from machines or intravenous apparatus to patients. The male and female components of Luer connectors join together to create secure yet easily detachable leak-proof connections. Multiple connections between medical devices and tubing are common in patient care. Luer fittings, connectors, and locks can easily connect many medical devices, components, and accessories. Because they are so easy to use, clinicians may mistakenly connect the wrong devices, causing medication or other fluids to be delivered to the wrong destination. These errors can cause serious injuries—even death. The Institute for Safe Medication Practices reports on multiple cases of fitting errors and cites specific examples of tubing from a portable blood pressure monitoring device inadvertently connected to a patient's IV line and the connection of oxygen tubing to a pediatric patient's IV line [44]. In 2008, the U.S. Food and Drug Administration published a set of case studies intended to sensitize clinicians to the problem [45]. An opportunity exists to improve the safety of interconnecting medical tubing. The quality methods associated with Poka-Yoke error proofing techniques [46] would appear to be applicable to an investigation in this area.

Remain flexible and be aware of the role that serendipity plays. That is, the team may move in one direction and discover something else which may be equally important or more important. The team should try to keep an open mind, to detect and understand the importance of the unforeseen incident and use it constructively. The children's toy Silly Putty was discovered when James Wright, a worker for General Electric, was told to create an inexpensive alternative for rubber. Most people know the story of biologist Sir Alexander Fleming who was researching a strain of bacteria called staphylococci. After returning from vacation, he noticed that staphylococcus bacteria seemed unable to grow in the area surrounding a fungal mold in one of the glass culture dishes—thus the discovery of penicillin. Spencer Silver was working in the 3M research laboratories trying to develop strong glue. His work resulted instead in an adhesive that wasn't very sticky. The Post-it note was born. Today, it is a popular office product because of serendipity.

Think about improving an existing product, process, or service by using the Failure Mode and Effects Analysis described in the appendix.

A word of caution. Don't think that complicated means better. And complicated may not necessarily be more difficult. Keep these words in mind from Steve Jobs:

> "That's been one of my mantras – focus and simplicity. Simple can be harder than complex: You have to work hard to get your thinking clean to make it simple. But it's worth it in the end, because once you get there, you can move mountains." [47]

SWOT Tool

A *S*trength, *W*eakness, *O*pportunity, *T*hreat (SWOT) analysis has been extensively used in business as an analytical, problem-solving, and project-planning tool. While the SWOT method was originally developed as a strategic planning tool for business and industry, it is equally useful in the work of healthcare, education, and even personal growth. The analysis can help a team focus and reflect on key issues and can bring these key issues to the forefront for a complete examination.

Consider performing a SWOT analysis using the questions raised in Table 4.1 as examples to help the team evaluate a product or service. A SWOT analysis may guide the team to identify the positive and negative attributes within the organization or team (S-W) and outside of it, in the external environment (O-T). The analysis can give the team a snapshot of the viability of the idea. Developing a full awareness of the situation can help with strategic planning, problem solving, and decision-making. The approach may help the team explore possibilities for the product or service that it may be considering. It may help the team identify an opportunity and clarify the team's direction.

While a SWOT analysis is typically conducted using the matrix shown in Table 4.1, the team could also just make lists for each category. Use the method that makes it easiest to organize and understand the results. As the team works through each category, don't be concerned about too many details. Use *bullet points*. Just capture the factors that may be relevant to each of the four areas.

Perform a SWOT analysis in the same way that the team conducted a brainstorming session. Select a comfortable setting with a variety of stakeholders present. A comprehensive and insightful analysis can be helpful in determining the team's direction.

Once the team is finished brainstorming, create a final, prioritized version of the SWOT analysis, listing the factors in each category in order from highest priority at the top to lowest priority at the bottom. The SWOT analysis will highlight possible gaps to help the team understand where it is, where it wants to be and how it will get there. The gap analysis may highlight deficiencies that may exist, which the team needs to resolve such as

- Labor: The needed skills and people resources.
- Methods: The processes needed to accomplish the effort.
- Metrics: The measurements needed to test and evaluate the unit.
- Equipment: The tools and apparatus needed for development and testing.
- Materials: The material items required to fabricate the unit.
- Time: Is the overall schedule realistic?
- Product or service functionality or features

Table 4.1 Typical SWOT analysis factors for a capstone team's idea

	Positive attributes	Negative attributes
Internal variables and present status	*Strengths* What are the team member's competencies and past experiences that bear on the project? What are the team's strengths? What makes the idea different or unique? What knowledge, skills, and attitudes do team members have that can make this project succeed? Will the team be using a mature technology for the project? Does the team have adequate resources (money, time, skills, facilities, equipment, etc.) to complete the project? Grants, funding agencies, venture capitalists, other sources of income exist to support the project. The team can offer a training program to potential clients and customers. The team is sensitive to customer needs. The team has a proprietary technology. Describe the intellectual property possibilities: • Patent. • Copyright. • Unique Trademark. The product or service is scalable.	*Weaknesses* What are the idea's weaknesses or drawbacks? What are the team's weaknesses? Is the team missing essential skills or competencies? Describe them? Describe the areas in which existing product or service competitors have overwhelming edge. Does intellectual property that others control have a negative impact on the viability of the idea? Does the product or service use sole sourced items? Is this a problem? Does the cost of entry into the product or service area prevent the team from proceeding? Does the team have a way to market the concept so that there is clarity in achieving a revenue stream and profit path? Does the lack of a track record hamper the effort? In what way? Are there a limited number of clients for this product or service? Will the team be using state of the art or legacy technology which could prove to be a problem? Are their supply chain issues?
External variables and future status	*Opportunities* What opportunities are open to the team? Are there products or services that have not yet been provided by the competitors? What commercialization options exist? What emerging market trends can the team take advantage of? How can the team turn strengths into opportunities? Potential future funding sources include college or university involvement, business incubators, foundations, donors, government, and venture capitalists Demographics—changes in the population's age, race, gender, culture may have a positive impact on the product or service. Useful opportunities can come from such things as: • Changes in technology and markets • Changes in government policy • Changes in social patterns, population profiles, lifestyle changes, etc. • Market growth • Global expansion • Legislation • Growing green movement	*Threats* What is the competition doing? Do others possess intellectual property that could be a problem? Do competitors enjoy a good reputation in the market? Are the requirements for your proposed product or service changing due to an environment that is in flux? Do industry policies affect the product or service? Do new federal/state requirements make the job harder… or easier… or necessary? Demographics—changes in the population's age, race, gender, culture may have a negative impact on the product or service. Are any of the following obstacles? If so, describe. • National or local economic conditions. • Rising material, production, and postage costs. • Changing needs and tastes. • Technology. • Vulnerability to the business cycle. • Government policies.

- Whether the effort will be a preproduction prototype or a proof-of-principle (POP) effort.
- An initial examination of the project's profit viability and return on investment potential

The discussion then centers on how the team will resolve or fill the gaps. The gap analysis should answer the following questions:

1. What labor, material, facilities, equipment, knowledge areas etc. is missing and needed?
2. What is required to mitigate weaknesses?
3. What is required to defend against threats?
4. What can be done to take advantage of the opportunities?

The analyses promote a better understanding of the barriers that must be overcome to complete the project.

Weekly Progress Status Presentation: Project Search Phase

During the first few weeks in the capstone or senior project class, students make short presentations of project ideas. Develop and practice an "elevator" speech [48]. The "elevator pitch" is a concise, prepared talk that explains your ideas, clearly and succinctly in no more time that it takes an elevator to go from the first to the top floor. You are selling yourself and your ideas. You want to convince two or three other class members to join with you and spend the next semester or two working on this project. Prepare a 5-min presentation (with or without PowerPoint slides) describing the idea and the need. In the short talk, you must convince others of the feasibility of the idea; that it will be an enjoyable, useful, and viable challenge; and that you will be a comfortable person to work with during the next few months. Remember that a good idea need not be complicated. You don't have to present a big idea. A big idea may be difficult to successfully execute. A simple idea may be boring but it also may more workable than the complex idea.

The class members may not know one another. This is the beginning of Tuckman's forming stage. Everyone is sizing one another up. Display passion and energy about your idea. In almost all aspects of life, people skills—part of the engineer's complement of "soft skills"—have a significant impact on success. Most employers are likely to hire, retain, and promote persons who are dependable, resourceful, ethical, having effective communication skills, self-directed, willing to work and learn, and having a positive attitude [49]. Welcome people with all types of skills. A successful team needs people with technical, analytical, software, hardware, and business proficiencies.

After finding and selecting a preliminary idea for a project, the team must gain approval from the faculty mentor in order to proceed. The team must convince the faculty mentor that the project is worthwhile and worthy of continued effort. Once more prepare a sales pitch—again the "elevator speech." Yes, an engineer must be a sales person too. In a conference with the faculty mentor, get to the point. Don't go on and on and on and on. The faculty mentor may not understand the technology or the team's approach, so pretend you are briefing someone in a kindergarten class. Practice the "keep it simple silly" (KISS) principle. KISS is a rule that states that technical systems are best when they have simple designs rather than complex ones. KISS is not meant to imply silliness. On the contrary, it is usually associated with designs and ideas that may be misconstrued as silly because of their simplicity.

Win the faculty mentor's interest in the project. Welcome questions because that will be the team's signal to get more technical. If the team is trying to get the approval of a senior executive in a company, don't forget to talk about Total Cost of Ownership and Return on Investment. In industry, even if you have thought of a way to improve an existing product or service, it will only gain approval from senior management if it can be shown that the implementation of this idea saves the company money within 3-years or less. If team members don't know an answer to a question—don't try to snow anyone. Be truthful, and say "Good question. I don't have an answer at my fingertips, but I can get back to you within a day." Get back with an answer in whatever time (an hour, day, or a week) to which you commit or the team will lose credibility.

Proposal Preparation and Presentations

Initial Activities

Market Research and Business Viability

The team has selected an idea. The next step is to discover if it is an idea that warrants the team's time for the next few months. If the idea is to result in a sustainable business the team needs information related to marketing, pricing, and cost in addition to technical know-how. The team begins by performing basic market research. Use the SWOT analysis as a starting point for developing answers to the questions:

- Who are the target customers and what are their needs? How large is the market for the product or service?
- If the project involves a process change, then what resources will be required over what time period? How much money will the organization save?
- Who are the significant competitors, what are they providing, and what is the price?
- How does your team's product or service differ from the competition? Is the proposed product or service better than the competition or unique?
- Can the business charge enough to cover costs and turn a profit?
- Will the selling price be competitive?
- What are prospective customers willing to pay for this product or service?
- Will location give the business or team an advantage?
- Is the proposed method of distribution superior to that of the competition?
- What are the market trends? Understand the demographic and societal trends that influence this product or service.

Think about the different business models and revenue streams. Will the fledgling company fabricate the product or outsource it and become a distributor or will the team license the invention to someone else. If the team is considering a web business, will it charge end users a monthly subscription fee, or make the website free for the user but earn a commission for every "lead" or sale that it generates for

Table 4.2 Product comparison template

Product or service feature	The team's product or service	Product from company A	Product from company B	Level of importance to the customer
Reliability				
Throughput				
Speed				
Price				
Ease of use				
Ease of installation				
Health and safety				
Environmental impact				
Power consumption				
Operating temperature				
Size				
Weight				
Shape				
Color				
Software				
Warranty				
Availability of training				
Security				
Built-In test capability				
Multilanguage flexibility				
Multiple interfaces				
Configurable				
Customizable				
Other features				

another company, or sell advertising space? Does the team want to consider the "freemium" model, where a basic service is offered for free but customers pay for upgrades for premium service? Examples of the freemium model are Skype, which is a VoIP service that provides audio and video calls through the computer and LinkedIn, the professional networking tool.

Conduct a competitive analysis to make sure the team's product or service has attributes that are equal to or better than those offered by competitors. While the team should perform a feature-to-feature comparison, make certain that the team's product or service offers several unique features. Identify ways in which the product or service helps the customer get a job done.

Another way to examine a product or service is to use a product comparison template that assesses product or service features based on importance to the customer as shown in Table 4.2. Add product features to this table as appropriate.

Highlight differentiating features in Table 4.2 that are important such as reliability or throughput. Even if the price of your product is higher than the competitor's, the novel features that the team's product offers may more than offset a higher price.

Discover and identify competitors by researching the market share that they command, the amount of publicity they receive, the number of mentions made by prospective customers, and the cost of their solution. Publicly held companies will have more information available than privately held companies. Information about companies can be obtained from product reviews, press releases, the company website, the company profile at www.linkedin.com, and reports from Dun and Bradstreet (www.dnb.com).

Whether the team intends to convince the vice president for research and development in an organization or convince a venture capitalist to invest in the product or service, the team must show that the idea will provide a reasonable return on investment. The executive in a company expects to recover the organization's investment in 2 or 3 years. If the team intends to start a new company, venture capitalists may offer a business a combination of an equity stake with a loan and might expect a return of between 10 and 20 times their investment within 5 years. The team needs to provide an analysis of the profit potential of the idea.

Intellectual Property

The team will have to consider the question of intellectual property (patents, licenses, trademarks, copyrights) issues. Are their intellectual property issues that should be considered? Chapter 7 discusses intellectual property in more detail.

The Basics

During this time, the team will begin to prepare an overall approach with a functional block diagram. The team will take into consideration the societal, environmental, political, ethical, health, and safety aspects of the project. The team relates client or customer needs to a project specification. Not only must team members agree on the project's features but they must also agree on what not to do. In part, the decision of what to include and what not to include in the effort will be governed by the technical skills available to the team as well as the need to stay within a prescribed budget and time period. During this phase, the team will provide weekly status review presentations and submit a brief log of their activities. All of this culminates in a proposal presentation and report.

Consider formulating a team identity to develop an *esprit de corps*. Create a name for the project. The team may design a logo to go on its documentation and presentations.

Project Budget

Remember: Whoever Controls the Finances Controls the Organization

Money is the grease that makes things happen in for-profit or nonprofit organizations. Make sure the team is aware of the amount of money allocated to the project and develop a budget that is in agreement with the funding. The budget is an estimate and the team should state the assumptions on which it is based. To properly develop a sound project budget, the team should outline the project activities as well as the required equipment and material to be purchased and estimate their cost, while keeping in mind time constraints and other dependencies that may impact the budget.

A list of all of the tasks to be performed is commonly referred to as a Work Breakdown Structure (WBS). A list of typical capstone project tasks to be accomplished is shown in Tables 4.3 and 4.4. Table 4.3 identifies the tasks associated with

Table 4.3 Typical capstone project proposal tasks

WBS no.	Task
1	Assemble Team & Project Proposal Kick-Off
2	Identify a Team Project Topic
3	Identify & Analyze Similar Products and Services
4	Evaluate Business Viability
4.1	Complete the SWOT and GAP Analyses
4.2	Complete a Marketing Analysis
5	Review Intellectual Property Conflicts
6	Identify In-Scope and Out-of-Scope Features
7	Prepare a Hardware Specification
8	Prepare a Software Specification
9	Identify Items for Purchase
9.1	Software
9.2	Hardware
9.3	Equipment
9.4	Training
10	Purchase Items
11	Prepare the Test Plan
12	Prepare the Formal Written Proposal
13	Prepare the Formal Proposal Presentation
14	Submit Formal Written Proposal
15	Deliver Proposal Presentation

Table 4.4 Typical capstone project development tasks

WBS no.	Task
1	Project Start
2	Track Project Progress
2.1	Status Reviews
3	Hardware
3.1	Review Specification
3.2	Design
3.3	Design Review
4	Software
4.1	Review Specification
4.2	Design
4.3	Design Review
4.4	Code and Test Modules
5	Verify Material/Hardware/Software Procurement
6	Hardware Fabrication & Integration
7	Overall Hardware and Software Integration
8	Perform Acceptance Test using the Design Verification Matrix
9	Documentation
9.1	Prepare Final Report
9.2	Prepare Final Presentation
9.3	Review and Confirm Business Viability
10	Submit Final Report
11	Deliver Final Presentation and Perform a Demonstration

Table 4.5 Capstone project budget template

Item	Item description and/or part number	Vendor name	Vendor contact (name, address, phone, website)	Quantity	Cost per unit	Cost
Personnel						
Equipment						
• List Item(s)						
Components						
• List Item(s)						
Software						
• List Item(s)						
Web Services						
• List Item(s)						
Materials & Supplies						
• List Item(s)						
Computer Services						
• List Item(s)						
Construction						
Marketing						
Fees						
• Permits and Applications						
• Testing						
• Legal						
• Other						
Other Direct Costs (ODC)						
• Consultant(s)						
• Training						
Food						
Travel						
Total						

the *proposal* phase. Table 4.4 describes the tasks for the *development* phase—tasks to "do" the project. Both lists should include key performance indicators (KPI), which are measurable indicators used to report the progress of critical project success factors. KPIs are intermediate as well as final outcomes associated with an idea, product, service, process, or procedure. Consider these task lists as a starting point for the proposal effort and development project.

Use the project budget template in Table 4.5 as a guide to creating a budget. If the team is uncertain of the specifics of the item to be purchased or the vendor from whom the item will be purchased insert a placeholder until the team obtains more details. A major reason that projects fail in industry is that they do not develop a realistic budget.

Either enter a zero or delete the line item if the team does not have expenses for a line item (or category). Insert or substitute categories of expenses that the team will use for the project. A brief description of several line items in Table 4.5 is shown below:

Personnel	Since the team members are not getting paid, the personnel cost line will likely be zero. Place the budgeted cost of a sub-contractor or consultant in the Other Direct Costs (ODC) line item
Equipment	Frequently equipment (e.g. computers, cameras, furniture, building facilities, drills, machinery, and test apparatus) can be obtained from the institution or borrowed from team members at no cost
Components	The components line might include required chemicals, or electrical and mechanical parts
Software	Project software needs may include database tools, simulation tools, spreadsheets, word processing, mobile application tools, time-management and scheduling tools, and a host of open source tools, which may or may not be free. Make use of Dropbox, Google Docs, or Microsoft's OneDrive (formerly SkyDrive) for free online file storage and file sharing
Web services	For example, web page hosting and security tools. Free phone conference services are available for interaction with team members. Try freeconferencecalling.com or freeconferencecall.com, but remember that participants will have to pay long-distance charges. Of course, Skype may be used between two people
Materials and supplies	Stationery, consumables, building materials
Computer services	Information processing support, data storage
Construction	Welding, fabrication of mechanical structures
Marketing	Ads, graphic design, mailing
Fees	As required
Other direct costs (ODC)	
• Consultant(s)	Many types of consultant information can be received at very low or no cost. For example, use the local utility to obtain electric, gas, or water information. Many states have soil and water testing laboratories that residents may use at no cost
• Training	Training needed to learn to use a software or hardware tool
Travel	Gas, tolls, lodging, air/bus/train fare

A risk reserve is not included in this template. If the team feels uneasy about this, then add risk funding as a category, or add a percentage amount under each category to take into account risk reserve.

An understanding of project finance and return on investment is important if the team wishes to proceed to the next stage of project development. At the end of the proposal, the team must be able to convince a company's management or venture capitalists that there is value to the proposal. The decision to implement a new product or service or make a change to an organization's process is ultimately based on the prospects of increased revenue or cost savings. Learn to talk the language of the people that control the purse strings—which is $$$. For example, safety improvements must be translated into how much money will be returned to the organization within a certain timeframe. Training must be shown to increase skills or other competencies, which will improve employee productivity or system efficiency. The purchase of new equipment must bring with it the likelihood of added revenue and improved profits. A manager will invest in a new or improved product or service or process change only if there is a resultant revenue increase within a relatively short time (typically 2–3 years). Inability to *demonstrate that an idea may deliver a financial return* will quickly turn administration optimism to pessimism and the budget will just as quickly disappear.

Weekly Progress Status Presentation: Project Proposal Phase

"Can you give me an update on the project?" That's something that all technical personnel hear and fear from the project manager. It's important to keep project stakeholders updated. This section discusses the preparation of great status reports—whether needed for an in-class presentation or for a boss in industry.

After assembling a team and selecting a preliminary topic, the team submits a weekly status progress log and presents a status update. It is likely that as time goes by, the project topic will morph into something quite different from where the team started. The redirection may occur for a host of reasons: the team's interests may change; the competitive study will indicate that other products or services exist that provide the same or similar features; a patent search reveals a conflict; the project scope was too ambitious; or simply the project was too difficult to complete within the time allotted.

On the other hand, if everything is on schedule and wonderful, that is great! Use status reports to share the good news. Everyone feels good when the team reassures people that they have everything under control.

Weekly Project Status Log

The project status log (Table 4.6) represents a short report that the team submits to the faculty mentor weekly. It is not as detailed as the weekly presentation. It provides a history of the team's activities and confirms that the team met regularly.

The significant outputs from the meeting minutes are who attended the meeting, who was assigned tasks, what were the tasks, and a brief tentative plan for work to be accomplished in the coming week. The weekly in-class status presentation requires more detailed information.

Contents of the Weekly In-class Project Status Presentation

The weekly project status tracking and reporting presentation is helpful for team members and the faculty mentor. It keeps everyone focused on what's done, what's not done, and what issues need to be worked. Although not cast in concrete, the weekly project status typically contains the following slides:
1) Project title, team member's names, and presentation date
2) Market Research and Business Viability Update
3) Budget Status Update
4) Planned Schedule
5) Week's Accomplishments

Table 4.6 Capstone project weekly progress status log

Project name: EPALS		
Instructor(s) Professor Hoffman	Class: CP 551	
Meeting Log for Week of: 11/17/14		
Type of meeting(s): Face to Face ☑	Conference Call ☐	NetMeeting ☐
Present at Meeting	Alice	
	Tom	
	Steve	
	Jane	
Accomplishments & Action Items:		
11/17/14 Meeting Minutes		
1. Group drafted a response to faculty member remarks concerning project ideas		
2. Response to instructor was e-mailed		
3. Drafted Weekly Status slides		
4. Jane and Tom researched available technology to establish a web service account		
5. Group worked on merging the WBS' created by individual group members into a single WBS		
Assigned Action Items (by team member)		
1. Create 1st draft of specification (Alice)		
2. Technology research of available electronics to realize remote activation device (Jane)		
3. Technology research of available SW to realize call center functionality (Tom)		
4. Create draft Bill of Material for a purchase order. (Tom)		
5. Refine WBS/schedule by assigning realistic timing for each task (Steve)		
Tentative plan for the meeting next week:		
1. Review draft requirements specification		
2. Review and further refine WBS/schedule		
3. Review and finalize Bill of Material		
4. Review software and hardware choices		
Next Meeting Date	11/24/14	Location: MCA 102

6) Issues
   a) Technical
   b) Schedule
   c) Budget and Financial
   d) Purchased material status
7) Project Risk
   a) Risk Matrix
   b) Risk Severity
   c) Risk Mitigation Plan
8) Plans for Current Week

Keep in mind that whenever a problem arises, the team must be able to respond with answers to the action questions:

What must be done?
When should it be completed?
Who will do it?
What is the impact to the schedule and budget?

> **Weekly Status Report**
> 1) Project title, team member's names, and date
> 2) Market Research and Business Viability Update
> 3) Budget Status Update
> 4) Planned Schedule
> 5) Week's Accomplishments
> 6) Issues
>   a) Technical
>   b) Schedule
>   c) Budget and Financial
>   d) Purchased material status
> 7) Project Risk
>   a) Risk Matrix
>   b) Risk Severity
>   c) Risk Mitigation Plan
> 8) Plans for Current Week

Market Research and Business Viability Update

As the team learns more about the selected idea or invention, they will be able to refine the need, the customers, the competition, and the price that can be charged for the product or service. Market research will help the team discover information about the current and future trends that can affect the business. Research helps the team become an expert about every aspect of the business area, including market size, the targeted demographic, and the market conditions that may affect your business. The more the team knows and understands these factors, the greater the possibilities for success. Market research is key to understanding the viability of the product, service, or process.

> **Weekly Status Report**
> 1) Project title, team member's names, and date
> 2) Market Research and Business Viability Update
> 3) Budget Status Update
> 4) Planned Schedule
> 5) Week's Accomplishments
> 6) Issues
>   a) Technical
>   b) Schedule
>   c) Budget and Financial
>   d) Purchased material status
> 7) Project Risk
>   a) Risk Matrix
>   b) Risk Severity
>   c) Risk Mitigation Plan
> 8) Plans for Current Week

Budget Status Update

The team should keep track of the expenses on a weekly basis as shown in Table 4.7. The planned budget is compared with the actual expenditures-to-date for each category or line item and the variance is shown in the last column. The team must monitor and evaluate the variances. If the project has deviated significantly (meaning 10 % or more) from the plan then a "workaround" is in order. The replanning effort may involve a reduction in scope, a change to the specification, or a modification of the items to be purchased. Expect and be ready to explain the variance. If a project doesn't show any variance, someone may be "cooking" the books.

Table 4.7 Capstone project budget status update—variance analysis

Date:			
		Actual expenditures week	
Expense	Budgeted amount	1 2 3 4 5 6 7 8 9 10 12 13 14 15	Expenditures to-date	Variance
Personnel				
Equipment				
• List Item(s)				
Components				
• List Item(s)				
Software				
• List Item(s)				
Web Services				
• List Item(s)				
Materials & Supplies				
• List Item(s)				
Computer Services				
• List Item(s)				
Construction				
Marketing				
Fees				
• Permits and Applications				
• Testing				
• Legal				
• Other				
Other Direct Costs (ODC)				
• Consultant(s)				
• Training				
Food				
Travel				
Total				

Weekly Status Report
1) Project title, team member's names, and date
2) Market Research and Business Viability Update
3) Budget Status Update
4) Planned Schedule
5) Week's Accomplishments
6) Issues
   a) Technical
   b) Schedule
   c) Budget and Financial
   d) Purchased material status
7) Project Risk
   a) Risk Matrix
   b) Risk Severity
   c) Risk Mitigation Plan
8) Plans for Current Week

Planned Schedule

Create a schedule using the tasks listed in the team's work breakdown structure or use Table 4.4 as a starting point. Every project's inextricably linked constraints are Scope, Schedule, and Budget. Changing one will impact the other two.

The presentation slide dealing with the schedule informs stakeholders where the team should be on the date of the presentation. The schedule could be a Gantt chart extracted from Microsoft Project (as a cut and paste item) or it can be simply a list of tasks with the associated start and finish date, and the person assigned as shown in the sample schedule shown in Table 4.8.

The faculty mentor is most interested in the progress against significant tasks and milestones. In particular, the faculty mentor may wish to view the schedule's critical path. The critical path is a series of tasks that must be completed on time for a project to finish on schedule. If any of those tasks are delayed, the project is in trouble. If a delayed activity is not on the critical path, then the activity delay may be a budget problem, but may not cause a delay in project completion.

There are several questions that need to be addressed during the weekly presentation:
1. Did the team complete the tasks due since the last report?
2. Are there any issues that need to be addressed now?
3. Which tasks are taking more time than estimated? Less time?
4. If a task is late, what is the effect on subsequent tasks?
5. Are any project team members over-allocated or under-allocated?
6. What tasks are due before the next report?
7. Has anything happened that will impact the final delivery date?

Proposal Preparation and Presentations

Table 4.8 Typical capstone project proposal work breakdown structure (WBS) and schedule

Date					
Task		Start date	Finish date	Percent complete	Assigned to
1	Assemble Team & Project Kick-off				
2	Identify a Team Project Topic				
3	Identify & Analyze Similar Products and Services				
4	Evaluate Business Viability				
4.1	SWOT Analysis				
4.2	Complete a Marketing Analysis				
5	Review Intellectual Property Conflicts				
6	Identify In-Scope and Out-of-Scope Features				
7	Prepare a Hardware Specification				
8	Prepare a Software Specification				
9	Identify Items for Purchase				
9.1	Software				
9.2	Hardware				
9.3	Equipment				
9.4	Training				
10	Purchase Items				
11	Prepare the Test Plan				
12	Prepare the Formal Written Proposal				
13	Prepare the Formal Proposal Presentation				
14	Submit Formal Written Proposal				
15	Deliver Proposal Presentation				

Table 4.8 includes a column that requires the team to indicate the progress for each task listed in the WBS as of a given date. Measuring a task progress using percentage completion is almost impossible. Can a team really measure the difference between 25 and 33 % complete? Instead use the column simply as an indicator that the task has not started (0 %), has begun (50 %) or is complete (100 %). The 50 % indicator is not really a measurement of the amount of work completed. It represents an acknowledgement that the work is in progress. Every task item should be no more than a week in length and should have a measurable output to enable confirmation that the task is complete. Violating this rule makes tracking the task difficult and neither the faculty mentor nor the team members may be able to determine if the task item is on schedule.

If task progress appears unsatisfactory, the faculty mentor will drill down into the details to discover the problems. The weekly status report needs to list the tasks due since the last report (including anything outstanding from the last report) and identify if they were met or not. If there is a problem, it must be identified and explained in this report.

```
Weekly Status Report
1) Project title, team member's names, and date
2) Market Research and Business Viability Update
3) Budget Status Update
4) Planned Schedule
5) Week's Accomplishments
6) Issues
   a) Technical
   b) Schedule
   c) Budget and Financial
   d) Purchased material status
7) Project Risk
   a) Risk Matrix
   b) Risk Severity
   c) Risk Mitigation Plan
8) Plans for Current Week
```

Week's Accomplishments

This should be almost identical to the "Plans for the Current Week" section from the previous status report. If a task was not accomplished then it must be addressed in the Issues section or the Risk slide.

The team uses this section to remind themselves of planned activities. This promotes a concentration of focus and a sense of progress as each planned activity on the list is initiated and completed.

Issues

```
Weekly Status Report
1) Project title, team member's names, and date
2) Market Research and Business Viability Update
3) Budget Status Update
4) Planned Schedule
5) Week's Accomplishments
6) Issues
   a) Technical
   b) Schedule
   c) Budget and Financial
   d) Purchased material status
7) Project Risk
   a) Risk Matrix
   b) Risk Severity
   c) Risk Mitigation Plan
8) Plans for Current Week
```

An issue is an event or condition that has *already* happened and has impacted or is currently impacting the project objectives. An issue *must* be resolved or it *will* adversely affect the schedule, budget, or technical performance. The language of risk is normally in the future tense whereas language of an issue is in the present or past tense. For example, a risk is that a critical human resource may leave during the project; whereas an issue is that a team member has resigned and is leaving on Friday

The Issues page is a place for the team to alert everyone to current problems that may delay or derail the project. The issue may come from the technical, schedule, or financial realms. It could be that a new unforeseen task has been identified. Perhaps a vendor has difficulty in shipping an item by the need-date; or a company no longer manufacturers a particular piece of equipment that the team ordered; or an important test in the test plan failed. If any of these situations occur, then the team must develop an alternative approach.

Risk

Weekly Status Report
1) Project title, team member's names, and date
2) Market Research and Business Viability Update
3) Budget Status Update
4) Planned Schedule
5) Week's Accomplishments
6) Issues
   a) Technical
   b) Schedule
   c) Budget and Financial
   d) Purchased material status
7) Project Risk
   a) Risk Matrix
   b) Risk Severity
   c) Risk Mitigation Plan
8) Plans for Current Week

Risk is a measure of uncertainty in achieving program performance objectives within the defined cost, schedule, and performance constraints. It has two main components:
1. The probability or likelihood of failing to achieve a particular outcome, and
2. The consequences or impacts of failing to achieve that outcome.

No matter how carefully a team prepares for all contingencies, inevitably the team will confront unexpected challenges. It seems obvious that if too much project risk exists, the project is in jeopardy. So both the team members and the faculty mentor want to know about risk. The team should identify risks by answering the

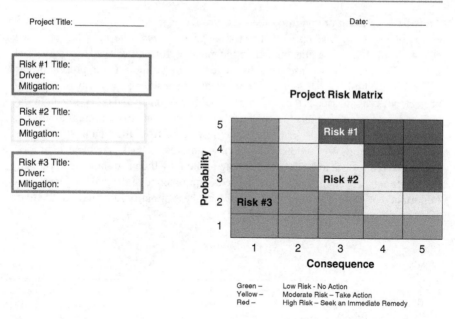

Fig. 4.1 Risk matrix

question "What can go wrong?" When considering risk, the team should take into consideration the available labor versus the work to be accomplished, the design complexity, supplier issues, resource availability, budget, etc. Carefully monitor software and hardware test results. Understand the reasons for any anomalies or discrepancies and be prepared to discuss them in class.

Go into a risk mitigation planning mode when the team discovers a problem. The team has to answer the *action* questions:
What should be done?
When should it be completed?
Who will do it?
What is the impact to the schedule and budget?

Figure 4.1 is a tool that is frequently used to graphically show risk. Red, yellow, and green offer a qualitative perspective on risk severity with red signifying danger and green signifying "forget about it" for the time being. Figure 4.2 quantifies and prioritizes the identified risk using risk severity as a guide. Assigning the probability score is tricky at best. This is based on subjective judgment. Use Table 4.9 to select the score that corresponds to the chances of occurrence.

Consequence or impact can be assessed in terms of its effect on cost, schedule, and technical performance as shown in Table 4.10. Choose the most appropriate domain for the identified risk from the left hand side of the table. Then review the columns in the same row to assess the severity of the risk on a scale of 1–5 to determine the impact score, which is the number given at the top of the column. A task may be impacted by more than one factor. If so, choose the one with the highest score.

Proposal Preparation and Presentations

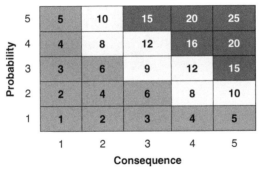

Severity = Probability * Consequence

Fig. 4.2 Converting the risk matrix to a severity priority list

Table 4.9 Risk matrix—probability score

Probability score	1	2	3	4	5
Descriptor	Highly unlikely	Unlikely	Possible	Likely	Almost certain
Chance of occurrence, %	<5	Between 5 and 20	Between 20 and 50	Between 50 and 80	Between 80 and 100

Table 4.10 Risk matrix—impact score

	1	2	3	4	5
Domain	Negligible	Minor	Moderate	Major	Catastrophic
Budget	Small loss or insignificant cost increase	<5 % over project budget	5–10 % over project budget	10–25 % over project budget	>25 % over project budget
Schedule	Slight slippage	Slight slippage against key milestones	Delay affects project stakeholders	Failure to meet deadlines in priority outcomes	Failure to meet primary objectives
Technical performance—compromised specification	Barely noticeable reduction in scope or performance	Minor reduction in scope or performance	Reduction in scope or performance	Failure to meet a significant objective	Failure to meet two or more primary objectives

Table 4.11 Severity calculation

Task no.	Brief task or problem description	Probability or likelihood	Impact or consequence	Severity
1				
2				
3				
4				
5				

Severity = Probability (1–5) × Consequence (1–5)

Table 4.12 Risk severity score

Severity score	Risk level	Color	Recommended response
15–25	High risk	Red	Immediate action. Prepare detailed plans for a work-around. Implement plans
8–14	Moderate risk	Yellow	Prepare action plans. Monitor situation closely
1–7	Low risk	Green	Limited or no action

Severity = Probability Score × Impact Score

The risk matrix requires the team to explicitly define potential risks and assign a probability and a consequence. The priority of addressing risks is based on the severity calculation, which is the product of impact/consequence and the probability/likelihood. Insert the results into Table 4.11, which organizes the risks based on the severity calculation. The calculation categorizes severity into red (15–25), yellow (8–14), or green (1–7) as shown in Table 4.12 and Fig. 4.2. Address the highest severity scores first. This order provides the priority in which the team should work. If the team identifies more than four high risk problems, begin mitigation immediately and consider alternative approaches.

The team then must take action in one of the following three ways:
1. Avoid Risk: The project takes action to avoid the risk. Typically this means the project chooses not to undertake an activity.
2. Reduce Risk: The project takes action to reduce either the probability or consequence of the risk.
3. Retain Risk: The project team recognizes that unanticipated events will occur and will choose to retain those risks that are of low consequence. The team may seek a method of circumventing or overcoming the problem (usually referred to as a workaround) or put additional manpower on the activity.

There is widespread use of the risk matrix in business and industry and variations of these charts exist. The risk matrix is one of the standard tools used to inform management about risk situations. It quickly adds understanding of the situation, even with the limitations associated with the variation in the assignment of probability and impact scores.

> Weekly Status Report
> 1) Project title, team member's names, and date
> 2) Market Research and Business Viability Update
> 3) Budget Status Update
> 4) Planned Schedule
> 5) Week's Accomplishments
> 6) Issues
>   a) Technical
>   b) Schedule
>   c) Budget and Financial
>   d) Purchased material status
> 7) Project Risks
>   a) Risk Matrix
>   b) Risk Severity
>   c) Risk Mitigation Plan
> 8) Plans for Current Week

Plans for the Current Week

This should be a bullet list of fewer than five items that the team hopes to review or accomplish in the next week. The plans for the current week not only reflects the overall project schedule but also a reaction to the risks and issues that arose during the week. The schedule represents a top level view of tasks that must be completed to move the project along. The plans for the current week represent a subset of the broader schedule. As a consequence of discoveries made during the week, the overall schedule may have to change. The week's objectives help the team to focus on specific details that must be done and identifies a person to do it.

Preliminary Design Review (PDR) and Critical Design Review (CDR)

Customarily, industry and government organizations conduct both a Preliminary Design Review (PDR) and a Critical Design Review (CDR). The PDR is a formal technical review of the basic design approach. It is held after completion of the software and hardware specification(s), the software design and test plan, the top level hardware design and the lists of required material and equipment but prior to the start of detailed design. The overall project risks associated with the technical, cost, and schedule are also reviewed.

Many capstone courses conduct a collective PDR that is spread over several weeks and do not have a single specific date for a formal PDR. On the other hand, a

CDR represents the culmination of the initial design effort. This section focuses on preparations for the CDR, which is conducted at the end of the first semester of a two-semester capstone course. Stakeholders attend the CDR to:
1. Confirm that the team has a grasp on the need for the project and possesses the skills and abilities to complete the project
2. Review the software and hardware product, service or process specification
3. Confirm as best as possible that the initial design satisfies the performance requirements described in the development specification
4. Assess risk with respect to technical, cost, and schedule

The CDR consists of two parts—a written proposal report (which includes a specification) and a presentation given at a meeting attended by the stakeholders, which is based on the proposal report. A successful review is a confirmation that the stakeholders have confidence that the team is ready to proceed to the next step. The stakeholders' approval signifies agreement that the team's design can likely meet the stated performance requirements within technical, budget, schedule, risk, and other constraints. The presentation represents a summary level view of the material in the written report. Once the presentation and the report are approved the design and implementation proceeds and the budget for the material to be purchased is released.

Project Specification

After the team has agreed upon a design project, it must decide on the details of what to do. The project specification represents a set of design requirements. Starting a design without an agreed list of requirements is foolhardy and will lead to schedule delays, budget overruns, and significant misunderstanding between the stakeholders (including the faculty mentor) and the team. Creating a well-defined specification helps to ensure that
- Designers know exactly what to develop
- The team knows what to test (this may be done by the quality assurance personnel in industry)
- The client or stakeholders know exactly what they will receive

Specifications are quantitative, measurable criteria that the product or service is designed to satisfy. A specification serves as the basis for determining the features of the project. The specification may include key performance indicators (KPI) which are measurable indicators used to report progress that reflects on the critical project success factors.

This is the time to decide on how the desired features determine or constrain size, velocity, response time, cost, weight, etc. With consideration of the potential users, the team thinks about cost, safety, user-friendliness, performance, compatibility with other products or services, functionality, acceptance, convenience, capacity, misuses, legal issues, standards or codes, reliability, availability, maintainability, materials, portability, productivity enhancement, entertainment, technology, manufacturability, societal factors (human interface, environmental factors), and design

Table 4.13 Criteria for a good requirement

1.	Necessary	If the system can meet the intended use without the requirement, it is not necessary
2.	Feasible	The requirement can be accomplished within the project's budget and time frame
3.	Concise	The requirement is stated simply
4.	Unambiguous	The requirement can be interpreted in only one way
5.	Correct	The facts related to the requirement are accurate and it is technically and legally possible
6.	Complete	All conditions under which the requirement applies are stated, and the requirement expresses a whole idea or statement
7.	Consistent	The requirement is not in conflict with other requirements
8.	Verifiable	Implementation of the requirement in the system can be proved
9.	Design independent	The requirement does not pose a specific implementation solution
10.	Nonredundant	The requirement is not a duplicate requirement
11.	No escape clauses	Requirements do not use if, when, but, except, unless, and although and do not include speculative or general terms such as usually, generally, often, normally, and typically

methods. The specification defines requirements in terms of inputs, outputs, throughput, data flow, and behavior of the system.

The specification requirements that the team develops should meet the criteria of a good requirement shown in shown in Table 4.13 [50]. If it does not meet the intent of each of the eleven criteria, it is likely that the requirement statement needs some work. As the team writes the requirements think about what constitutes project success. For each requirement the team must develop a test to verify that the product or service meets the requirement.

The specification is part of the proposal report submittal. A design should not commence without an approved specification. The specification template shown below should be used as a guide. Include the items that pertain to the project and omit those that are not applicable.

Specification Template
1. Project Title
   In addition to the project title, the title page should include a list of team members, the course number and name, and the date prepared.
2. General Information about the product or service or system
   1.1. Purpose
   1.2. Scope
   1.3. Product architecture and overall block diagram
3. Project Standards and Applicable Governmental Regulations
   A specification might include commercial standards, military standards, company in-house standards, or product design standards. Standards are created by a wide variety of organizations. The most common are professional societies, industrial or manufacturing associations, domestic or international

governmental agencies or bodies such as the U.S. Department of Defense, local town or city regulations, and companies. Company standards are often proprietary and therefore are available only to approved subcontractors. Typical standards that bear on an engineering design project may come from any of the following organizations:

AASHTO: American Association of State Highway and Transportation Officials
ACI: American Concrete Institute
ANSI: American National Standards Institute
API: American Petroleum Institute
ASCE: American Society of Civil Engineers
ASHRAE: American Society of Heating, Refrigerating, and Air-Conditioning Engineers
ASME: American Society of Mechanical Engineers
ASTM: American Society for Testing and Materials
AWS: American Welding Society
CE marking indicates compliance with the European Union's product quality legislation
CSI: Construction Specifications Institute
DIN: Deutsches Institut für Normung (German Institute for Standardization)
DOD: U.S. Department of Defense, U.S. military standards
DOE: Department of Energy
ESI: European Software Institute
FAR: Federal Acquisition Regulation
HIPAA: Health Insurance Portability and Accountability Act
IEEE: Institute of Electrical and Electronic Engineers
IESNA: Illuminating Engineering Society of North America
ISO: International Organization for Standardization
ITE: Transportation and Traffic Engineering Handbook, Institute of Transportation Engineers
JCAHO: Joint Commission on Accreditation of Healthcare Organizations
LANL: Los Alamos National Laboratory
LEED: Developed by the U.S. Green Building Council, Leadership in Energy and Environmental Design (LEED) is a rating system for the design, construction, operation, and maintenance of green buildings and homes.
NEC: National Electrical Code
NEMA: National Electrical Manufactures Association
NFPA: National Fire Protection Association
NIOSH: National Institute for Occupational Safety and Health
NIST: National Institute of Standards and Technology, agency of the U.S. Department of Commerce.
OSHA: U.S. Occupational Safety and Health Administration
SAE: Society of Automotive Engineers
SEI: Software Engineering Institute
UL: Underwriters Laboratories
USDOT: Manual on Uniform Traffic Control Devices for Streets and Highways, U.S. Department of Transportation, Federal Highway Administration

4. Assumptions and Dependencies
   Describe assumptions or dependencies regarding the hardware and software and its use that apply. These may concern such issues as equipment usage, or software and/or hardware interfaces.
5. General Constraints
   Describe limitations or constraints that have a significant impact on the design of the product or service or system's hardware and/or software (and describe the impact). Constraints may be imposed by any of the following:
   Hardware or software environment
   End-user environment
   Standards compliance
   Interoperability requirements
   Interface requirements
      Security requirements
      Hazardous material issues
      Performance requirements
      Communications
      Hardware and software integration issues
      Verification and validation requirements (testing)
6. Project Description
   Discuss any of the following topics that apply to the product, service, process, or system:
   Back-up and recovery requirements
   Data migration approach
   Capacity analysis
   IT functional analysis
   Data requirements
   Functional and performance requirement description
   Hardware and software description
   Input and output requirements
   Mechanical enclosure(s)
   Performance requirements
   Support considerations
   Systems and communication requirements
7. Attributes
   Describe product, service, or process attributes that apply. Be specific and quantitative. Consider verifiability, availability, portability, reliability, reusability, robustness, testability, efficiency, radio frequency interference and noise production, and usability.
8. Maintainability and Support Requirements
   8.1.1. Maintenance Requirements
   8.1.2. Supportability Requirements
   8.1.3. Adaptability Requirements: Identify components and procedures that are designed to be subject to change.

9. Timing
   Throughput time
   Sequential relationship of functions
   Priorities imposed by types of input
   Timing requirements
   Sequencing or interaction of activities within a system
   Input/output transfer time
10. Design Description: A description of the critical parameters that identifies what the product should be and should do.
    Algorithm development
    Architectural block diagrams
    Capacity
    Communications environment—speed, network, throughput, type
    Component selection
    Connectors and cabling
    Data flow diagrams
    Database schema—tables, fields, and relationships
    Functional and performance requirement description
    Housing and material
    Human–machine interfaces
    Inputs—electrical, analog, digital, mechanical, power, torque, data, sensors etc.
    Outputs—electrical, analog, digital, mechanical, power, torque, data, display etc.
    Power requirements
    Thermal requirements
    Printed circuit board
    Scalability
    Size, shape, weight, color
    Speed
    Shelf life
    Service life
    Use cycle
11. Interfaces
    11.1. User Interfaces
    11.2. Hardware Interfaces
    11.3. Software Interfaces
    11.4. Safety and/or regulatory compliance
12. Reliability—a measure of how well a product performs under a certain set of conditions for a specified amount of time. May be required to determine the length of a proposed warranty or prove that the product meets specific safety or regulatory requirements. Reliability is quantified as MTBF (Mean Time Between Failures) for repairable product and MTTF (Mean Time To Failure) for nonrepairable product.
13. Security
    13.1. Access Requirements
    13.2. Integrity Requirements

13.3. Privacy Requirements
13.4. Audit Requirements
14. Safety and Hazardous material Issues
15. Environment: Example environmental conditions include the following:
Operating and storage temperature levels
Operating and storage humidity levels
Operating noise level
Vibration levels
Shock loading
Exposure to dirt and other contaminants like salt spray, oil, or gas
16. Quality: If appropriate, discuss specific quality plans.
17. Supporting Documentation—if necessary
18. Testing: Procedures for testing the functionality and performance
19. Definitions, Acronyms and Abbreviations

Proposal Report

Organizations prepare a statement of work (SOW) prior to embarking on a new project. The SOW is a narrative that defines the scope of work and the time in which the effort is to be performed. A SOW typically includes these items:
- A list of major deliverables and when they are expected.
- Identification of the tasks that support the deliverable activities.
- The resources required including facilities and equipment and who provides the resources.
- A well-defined statement of the results to be achieved including quality and testing requirements.
- Quantifiable criteria that must be met for the work to be acceptable and accepted.
- Data requirements.
- Reporting requirements.
- Expected reviews and technical meetings.
- A schedule of milestones.
- A payment breakdown including who will pay for which costs and when.
- Installation and training requirements.

A business plan is a written description of a business's future, a document that informs interested parties about the plan for a product, service or process and how the plan will be implemented. In addition to describing a product, service, or process, it details the operational plans, marketing strategy, and lays out the sales forecast to keep the business on track for profits.

If outside investment or loans are sought, whether from venture capitalists or bankers or others, a business plan is essential. The business must show investors how they are going to make money. The business plan is one of the first documents that a loan officer will want to see. An angel or venture capital investor will want to not only see and read the plan before providing funding, but they'll try to poke holes in the business plan and ask questions about missing items that they believe should be addressed.

The proposal report described in this book is a cross between a statement of work and a business plan. The proposal report represents the team's written project plan. It consists of documents dealing with who, what, why, when, where, how, how much, and when will the business will make a profit or how much savings will accrue to the organization. The project proposal plan encapsulates much of the team's work accomplished at a midpoint in time, typically after the first semester of a two-term capstone class. The project proposal template on the following pages presents a format and content that covers much of the information needed by a business executive or venture capitalist before they give their approval to proceed with the effort. The ordering of the sections is not meant to imply the order to develop the sections. The order is meant for ease of reading, presentation, and use, and not as a guide to the order of preparation of the various sections. Tailor the project proposal plan to specific project and class needs. The final project report will be very similar to the proposal report that follows. The proposal represents intent to accomplish a set of items and final report includes the results of the effort.

The Capstone Course *Proposal* Report Format Template
1. Cover Sheet
   Title of Project
   Team Members
   Course Number, Course Title
   Date submitted
2. Abstract
   A short abstract (300 words maximum) should open the paper. The purposes of an abstract are:
   1. To give a clear indication of the objective, scope, and key results of the project so that readers may determine whether the full text will be of interest to them.
   2. To provide key words and phrases for indexing, abstracting, and retrieval purposes.

   The abstract should present the proposed problem, outline the approach taken, and present the key results or tasks to be accomplished. The abstract should not attempt to condense the whole subject matter into a few words for quick reading.
3. Description of the Project Problem
   Describe the problem or need that the team is addressing. Identify the purpose or objective of the project, the context of the project and the general technical problem (the type of project the team is doing (e.g., software prototype, hardware prototype, service, simulation, application program, etc.)).

   Describe the tasks to be performed. Include information about the research, analysis, hardware design, software design, and cost effectiveness (e.g., manufacturability). Identify required resources, major risks and provide a task schedule.

   At a minimum, provide a list of the major milestones. A milestone is a task of zero duration that shows an important achievement or project event. Milestones have zero duration because they symbolize an achievement at a point of time in a project. Major milestones represent points in the project where significant progress

Proposal Preparation and Presentations 73

has been achieved. In a project in industry milestones will be included that correlates with the payment plan in the Statement of Work. Milestones are frequently based either on the acceptance of formal deliverables, receipt of nonformal deliverables, or agreed upon progress updates. Progress updates may take the form of presentations, status reports, or meetings.

The team may use a table similar to the one shown below to list the major milestones that will be (or were) tracked during the project. Alternatively, the team may insert a copy of the Microsoft Project Gantt schedule that will be used.

Milestone	Date

4. Background

Discuss the context and history of the selected topic and describe what has been done in the past. Discuss the future direction of the product or service and how it might impact society. Consider addressing several of the following issues that may be of interest to the stakeholders:

Economic or financial: effect of this topic on the local economy, savings, possible cost of project development, material cost, labor issues, outsourcing needs, etc.

Environmental: influence on the environment in the past, possible impact of future developments

Sustainability: product life cycle

Manufacturability: material availability, use of commercial-off-the-shelf (COTS) versus custom components, special needs for hostile environments

Ethicality: uses that could cause harm to society, ethical issues that someone working on this topic might encounter

Health and safety: positive or negative impacts on the health and safety of individuals or society for past or future applications

Social: relationship of this topic to education, culture, communication, entertainment

Political: relationship of this topic to political issues

Include the results of a literature search for the overall problem. Examine the body of published information and data relevant to the team's topic. The team will identify and evaluate the information (including intellectual property) that is known about the specific subject area under investigation. The literature search should contain information from a variety of sources including books, periodicals, newspapers, journal articles, Web documents, intellectual property databases, and corporate publications.

Through the literature review the team will discover whether the problem area has been previously addressed and solved by another person, group, or company. If the search uncovers a prior solution, the effort changes to determine the efficacy of the existing solution and how complete it is. Perhaps modifications to an existing solution will make the product or service significantly more useful. The search may identify areas which have not yet been investigated or it might suggest a particular focus that was overlooked.

Use appropriate citations in the team's work. Citations document the sources that underpin concepts, positions, propositions, and arguments. Citations reveal the quality of work that supports a discussion. Always cite
- Statistics or specific numbers.
- Statements made by specific individuals, companies, or reports.
- Opinions made by others.

Readers may want either to verify information, or to learn more about issues and topics addressed. Citations demonstrate that a position or argument is thoroughly researched. Finally, the report should give proper attribution to those whose thoughts, words, and ideas are used.

5. Design Requirements or Project Specification

Specifications and requirements for the project—Include the detailed specification that serves as the basis for the project as an appendix. Include expected customer requirements and desired features related to form, fit, function, and interfaces. Consider aspects such as cost, safety, user-friendliness, performance, acceptance, convenience, capacity, misuses, legal issues, standards or codes, reliability, availability, maintainability, materials, productivity enhancement, entertainment, technology, and design methods.

Functional decomposition of the project—Explain the major functions required by the design. Use illustrations, drawings, and tables to supplement the discussion.

Selection of design criterion—Specify goals for performance, reliability, cost, memory requirements, manufacturability, safety, societal factors (human interface, environmental factors, etc.), and any other criteria relevant to the project.

6. Test

Describe what constitutes project success and why? Discuss the product or service tests that will confirm that the project succeeds in doing what it was intended to do. Use the design verification matrix described in Table 5.3 to describe the test results in the final report.

7. Business Aspects

Discuss the novel aspects of this service or product. Address why a company or investors should invest money in this product or service development. Although the team will likely be able to better answer the questions shown below at the end of the project, the team should do its best to address the following:
1. Describe the market and/or industry
2. What is the economic outlook for the industry?
3. Describe the novel features of this product or service.
4. Competitive landscape—what is out there? How does this product or service differ from that offered by other organizations?
5. Describe intellectual property or patent issues—if any? How would the team protect its intellectual property?
6. Who are the projected customers or clients?

7. Describe the resources needed to develop, test, evaluate, and market the product or service.
   8. What are the key competitive factors in this industry?
   9. How would the team produce or manufacture the product? Describe the distribution plan.
   10. Describe any needed testing and quality control procedures. Describe standard or accepted industry quality standards.
   11. What if anything would be subcontracted to other firms? Will a profit be made on the subcontracting effort?
   12. What is the strategy for gaining and/or keeping a client base?
   13. What is required to:
       Bring the product or service to market?
       Keep the company competitive?
   14. Describe the plan to keep the product or service on technology's cutting edge?

8. Final Implementation

If the project was a proof-of-principle discuss the realization of the final implementation. Describe the final implementations and its functions if different from the proof-of-principle (include diagrams, photos, code examples, and other illustrations in the body of the text and place large engineering drawings, listings, etc. in the appendices). Discuss and present the calculations used in the design of the project in the relevant subsections.

9. Deliverables

Describe the output or deliverables that the team will provide. Relate the deliverables to significant milestones during the project. Use Table 4.14 below as a guide in selecting deliverables.

10. Scope of Work to Be Excluded

Describe any items and tasks specifically excluded from the scope of this project, for example
- Preparation of user aids and training material
- User Training
- Any revision or development of standard business practices or procedures

By clearly stating tasks that will not be done, the team eliminates potential misunderstandings between customer and developer.

11. Performance Estimates and Test Results

State the estimated project performance criteria (and how they were derived). Estimates may include references to speed, cost, power consumption, noise-immunity, ease of use, etc., depending on the project. Refer to the specification if this material is described therein.

Discuss the initial performance results, if available. Compare the results with the predicted performance and explain discrepancies. Include suggestions for design changes that would improve the performance. Use graphs or other figures to show relationships when appropriate.

Describe the tests that the team will use to verify that the design or implementation does or does not satisfy the initial specification.

Table 4.14 List of project deliverables

Project stage	Deliverable	Description	Acceptance procedures
Concept definition	Conceptual architecture definition	Describes the technology and preliminary design. Includes a block diagram or schematic	
Planning	Basic plans	Work Breakdown Structure (WBS) Items to be purchased Key Performance Indicators (KPI) Schedule Hardware Specification Software Specification System Specification Test procedures	Preliminary design verification matrix
Design	Basic engineering design Bill of materials List of items to be purchased	Block Diagrams. Solution approach Detailed schematics Data flow diagrams Coding	
Project proposal and presentation	Proposal report and presentation to stakeholders		
Development	Design Risk mitigation plans Installation plan Data migration plan Training documentation	Software and hardware design activities. Contains identifiable outcomes to confirm progress Plans and results of installation and data migration activities, as appropriate	
Project test	Completed acceptance test	Use a design verification matrix that is based on the specification	This deliverable will follow the acceptance procedures or tests developed for this task Copy of the design verification matrix with test results
Report writing and final demonstration	Written report	Two major reports will be submitted. The proposal and the final report that describes the team's accomplishments Demonstration	View demonstration
Presentation and/or demonstration	Presentation to stakeholders and a demonstration of the product, service, or process		

12. Progress to Date

Describe the progress the group has made on proving the concept to a point in time. Include progress on selected key performance indicators.

13. Individual Contributions

Describe in detail what each person (by name) in the group will contribute or did contribute to the overall concept or idea the team is trying to prove and the role they played in the project effort.

14. Financial Considerations

The capstone course challenges engineering students and requires them to use critical thinking skills from different and possibly unfamiliar disciplines. Students are forced into learning new areas without traditional "spoon-feeding" that frequently accompanies classes. Stretch tasks challenge the team members' knowledge, skills, and creativity in new and different ways. The project's financial aspect requires understanding a different knowledge area, which is why it might be considered a stretch task.

This section consists of two parts. First the team updates and includes the project budget based on a version of Table 4.5. The team must also confirm that the faculty mentor has authorized the funding to accomplish all that has to be done.

Suppose the team decides at the conclusion of the project, to go forward with an entrepreneurial venture. All entrepreneurs have a host of decisions to make. Student social entrepreneurs go through a for-profit or nonprofit operation decision-making process. Many of the financial considerations are identical for a profit or nonprofit venture. Both for profit and nonprofits are appropriate for doing societal good. Both need funding to begin and move forward.

It takes a lot more than a good idea to develop a successful product or service venture. The start-up team needs to know where to find both financial and technological resources. The team needs to find the right people with the right skills to do the job. At some point the entrepreneurial team may consider pursuing financing for the project. Funding for the new venture may be negotiated by assembling a consortium of investors, lenders, and other participants to underwrite the effort. Not only does the start-up need product development funding and the investment required to produce the product or service, but a whole host of other activities requires capital. These efforts might include environmental testing, sustainability development, design for manufacturability, market entry, and distribution issues, and solutions to ethical, social, and political questions that may arise.

Not long after the for-profit start-up business starts capital must be returned to the investors. The investors or lenders not only expect to recover their investment within a few years but they also want to make a substantial profit. They expect to see a proposal that shows them that the team understands the need to present a financial picture that supports the idea of generating a profit in a short time frame.

Consequently, the second part of this section of the proposal requires the team to use its imagination to develop a somewhat realistic financial plan taking into consideration the type of business and operation for which the product or service has been developed. The project proposal may contain an initial examination of an entrepreneurial project plan. The final project report may include a more detailed

Table 4.15 Multiyear business revenue and cost work sheet

	Operating year 1	Operating year 2	Operating year 3	Operating year 4
Baseline costs				
• Labor				
• Consulting fees				
• Material				
• Equipment				
• Manufacturing				
• Facility rent				
• Utilities (telephone, gas, electricity, etc.)				
• Repairs and maintenance				
• Postage				
• Advertising and marketing				
• Website/E-commerce				
• Supplies				
• Licenses and permits				
• Product distribution				
• Professional fees (accountant, attorney, etc.)				
• Insurance and liability				
• Training				
• Travel				
• Meals and entertainment				
• Other				
Subtotal				
Projected revenue				
Profit or loss (nonprofits may call this line Retained Earnings or Funds for Investment or Surplus Revenue or Change in Net Assets or Cash Reserves)				

business financial plan at which point the team will have gained a greater understanding of the product or service, customers' needs, and the business environment.

As part of the competitive analysis, the team needs to set a sales price for the product or service offered. One method of calculating the selling price takes into consideration the cost of producing the product or service, and then adds a percentage for overhead and profit to arrive at a selling price.

Overhead is a catchall phrase for costs incurred that are not directly related to getting the product or service to customers. Overhead relates to fabrication and assembly costs, administrative costs, electricity, office cleaning, heat, equipment depreciation, and support labor. In the corporate world, departments such as purchasing, contracts, legal, accounting, and executive management are added to overhead. The team then compares its target price with the competition's price and describes the value difference.

Table 4.15 can assist the team in the preparation of a multiyear revenue stream chart that summarizes the anticipated initial costs including overhead, revenue and profit, or savings potential. The team estimates the projected sales for the product or service for 3 or 4 years. A multiyear profit and loss projection assists the team in estimating the return on investment (ROI). This is very important. Private and public companies will invest money in a product or service development if they can recover their money and begin to make a profit within 3 years. A longer time frame than 3 years may doom the effort. In the present environment, ecologically beneficial or environmentally sustainable activities are frequently permitted a longer ROI timeframe.

Generally when a business chooses to undertake a project, profit from existing activities funds the effort. Management will fund projects in order to develop new products or services that they can profitably market. They might also invest funds to develop process changes to improve efficiency or productivity within the organization. Finally, they might invest in new equipment, training, software, or a tool from which they will derive cost savings. As with the investors in a start-up operation, a business expects to recover its money within a few years and then make a profit or enjoy the savings accruing from the investment. Management expects to see a financial picture that supports this idea before they risk the organization's money in a project.

As shown in Table 4.15, profit can be measured and calculated. Simply put,

$$\text{Profit} = \text{Total Sales} - \text{Total Costs}.$$

Profit from a business is the amount of revenue that exceeds the expenses, overhead costs and taxes needed to sustain the activity. When the equation has a positive result, for-profit organizations call the result profit and the number is an indicator of the firm's success. However, nonprofit organizations call the result Retained Earnings or Funds for Investment or Surplus Revenue or Change in Net Assets or Cash Reserves. For either organization a negative number spells trouble.

There are five funding categories available to nonprofit organizations to obtain support for projects. These are:
1. Public funding (local, state, and federal)
2. Foundation funding (grants)
3. Bank financing (bonds or commercial loans)
4. Cash reserves
5. Capital campaign

Unless the project is a very significant capital effort, the nonprofit organization will use its cash reserves to provide the funding required for a project that will improve its efficiency or productivity. As with the investors in a start-up operation or a business' management, the nonprofit organization expects to recover its money within a short time and then reap the savings on the endeavor. The nonprofit's management also expects to see a financial picture that supports this awareness before they risk the organization's money in a project.

Many college capstone projects involve service learning which has been integrated into the mainstream college curriculum. While service learning can be a relatively low-cost strategy for achieving a wide range of academic and community goals, there are costs. Here again the college or the community will be more supportive of a project that provides a financial benefit to the beneficiaries. The team should be able to demonstrate that the benefits outweigh the costs.

The project team must offer a method of demonstrating to these different constituencies that the organization's investment will prove to be valuable and constructive.

Initially the team will make intelligent guesses to complete Table 4.15. Some of the line items may not apply to the product, service, or process on which the team is working. Several line items are not needed to gain a company's in-house approval to proceed with a project. If the team is involved with a nonprofit organization in the development of an activity that benefits society, the operation must show that savings will accrue to the organization as a result of implementing the intended project. A positive money entry must be entered on the institution's accounting books. It takes more than passion and purpose to keep alive a budding organization.

15. Summary Feasibility Discussion

This feasibility discussion represents a brief summary justification for the overall approach. Will the idea work or not? A project has to be viable not only in technical terms but also in economic and commercial terms too. There always is a possibility that a project that is technically possible may not be economically viable. Project feasibility analysis is a multistep exercise. This section is not intended to be a comprehensive analysis. The feasibility discussion is intended to be a review of the facts to confirm that the project is worthy of proceeding.

The team must consider and comment on the following six activities and show that they lead to positive results:
1. Market analysis—the need exists
2. Technically possible—hardware, software, choices of methods
3. Competitive market analysis—the product, service, or process provides value
4. Financial analysis—the project can be completed within the assigned budget
5. Schedule—the project can be completed within the allotted timeframe.
6. Operational issues—confirm the willingness of the college or organization to support the proposed project.

16. Future Work

Future work associated with the project may include tasks deemed out-of-scope because of technical, schedule, or financial considerations. However, this section may be more appropriate for inclusion in the final report. Describe possible future research or development activities. What additional research or development could be done by subsequent teams in the area? The FMEA described in the appendix may be a tool to assist the team in developing ideas for improving the product, service, or process.

17. Discussion, Conclusions, Lessons Learned, and Recommendations

Describe the issues that arose during the planning and implementation phase and how the team addressed it. Discuss lessons learned from participating in this project to this point.

18. References

Provide a bibliography citing the references used for background work, the literature search, the specification of parts, cost comparisons, etc.

19. Appendices

Appendices may include supporting material such as explanatory documents and oversized drawings. Appendices may contain a reference section, matter that is tangentially related, or a summary of the raw data or details of the method behind the work. It may represent information that is not essential for clarifying a report statement or conclusion but does substantiate that statement or conclusion in a meaningful way. If the information is not directly relevant to a statement made in the team's report then place it into an appendix.

Proposal Presentation

The proposal presentation represents a brief summary of the material in the written proposal report. Both the written proposal report and the proposal presentation are based on the work that the team accomplished at a midpoint in time, typically after the first semester of a two-term capstone class. Use the 12 part guide shown below as the basis for material to be discussed in the proposal presentation.

1. Title slide with project team members and date
2. Background (Optional): Inform the audience about the history or driving force behind the need. The status of current products, services, processes, or industry offerings.
3. Problem Definition and Need: One to three slides showing the problems or limitations associated with current solutions and customer needs.
4. Project Description and Objectives: Describe the project.
   a) Explain the novel features.
   b) List the objectives that the team plans to achieve.
5. Project Deliverables
   a) Deliverables
6. Project Scope
   a) Define what is in and out of scope
7. Design Overview: One to three slides describing the technical design. Include a block diagram.
8. Test Methodology
   a) What constitutes project success and how will the team assess it?
   b) Discuss the development and use of a design verification matrix (Table 5.3).
9. Business analysis: Answer as appropriate
   a) Approximate required project budget
   b) List competing market products and/or patents or other intellectual property
      i) Who are the industry competitors and describe the differences between the selected project product, service, or process and theirs. Use a condensed version of Table 4.2 (Product Comparison Template).
      ii) Identify existing patents that perform a similar function and describe the differences between the selected project product, service, or process and theirs.
   c) Who is the anticipated customer or client base?
   d) What are the projected sales? How did the team arrive at this number?

e) How much will this product or service cost the customer or client? Include information related to the initial purchase, installation, and maintenance. Show the cost analysis—as best as possible.
f) What is the Return on Investment (ROI) to a potential user or investor? The team may use a simplified version of the multiyear revenue stream chart developed in the proposal report to demonstrate the return on investment.
10. Possible project risks and mitigation methods
11. Project Schedule
12. List key milestones completed

The written proposal and specification should be made available to the stakeholders at the meeting.

Project Development

The information presented to the stakeholders during the project development phase is almost identical to that offered during the proposal preparation phase of the project. Students continue to submit the weekly status log.

Weekly Progress Status Presentation: Project Development Progress Phase

The weekly progress status report is more critical in this phase than during the proposal phase. The team continues to meet weekly or more often. The focus is no longer on the business aspects of the project or competitive analysis or patent search. The specification is complete. The team is now in the "doing" mode.

Contents of the Weekly In-Class Project Status Presentation

The weekly project status tracking and reporting presentation is crucial to determine that the project is progressing satisfactorily. Again, the weekly project status typically contains the same topics as in the proposal phase:
1. Project title and team members
2. Schedule
3. Week's Accomplishments
4. Issues
    a) Technical
    b) Schedule
    c) Financial
    d) Ordering material
5. Project Risks
    a) Risk Matrix
    b) Risk Severity
    c) Risk Mitigation Plan

6. Plans for Current Week
   At each weekly meeting the faculty mentor will want to know
   - Did the team achieve the expected milestones?
   - What milestones are due before the next report?
   - Has anything happened that will impact the final delivery date?

   If a problem arises during the week, the team must answer the action questions:
   - What is the nature of the issues?
   - What must be done?
   - When should it be completed?
   - Who will do it?
   - What is the impact to the schedule and budget?

   If a risk is uncovered the team must answer the questions:
   - What is the risk?
   - What are the alternative approaches?
   - When should the team implement the alternative approaches?
   - Who will do it?
   - What is the impact to the schedule and budget?

Final Report, Presentation, and Demonstration

At this point in the semester, the team has completed the project development. The remaining effort surrounds the preparation of the final report and the final presentation.

Final Project Report

The final report focuses on project results and business conclusions and uses the proposal report as a starting point on which to build. Although it may not be welcomed, the team soon recognizes that change is inevitable. Update the information used in the project proposal sections resulting from modifications, corrections, and other adjustments made during the project development phase.

Address the reasons for changes and the methods used for evaluation and project control. Explain how the team controlled the damaging effects of "scope creep." Discuss the impact on the schedule, budget, and task efforts.

Provide the updated product or service specification as an appendix to this report. As shown below, the final report format is the same as used in the proposal with the changes included that result from the team's research and development efforts. Of course, tailor the final project report to the team's and class' needs.

The Capstone Course <u>Final</u> Report Format Template
1. **Cover Sheet**
2. **Abstract**

3. **Description of the Project Problem**
4. **Background**
5. **Design Requirements or Project Specification**
6. **Test**
   Summarize the test results in the design verification matrix and attach the details as an appendix.
7. **Business Aspects**
   Update the contents of the proposal response. Discuss reasons for changes.
8. **Final Implementation**
   Presentation of the final implementation. Provide updated calculations, block diagrams and/or new photos. If the team took a video of the project in action, post it online and provide a YouTube URL.
9. **Deliverables**
10. **Scope of Work excluded**
    Update the excluded work discussed in the project proposal that reflects the changes resulting from modifications, corrections, and other adjustments made during the project development phase.
11. **Performance Test Results**
    Discuss the results, compare with estimated performance and explain discrepancies. Include suggestions for design changes that would improve the performance of the project. If something didn't work as planned, then explain why.
12. **Progress**
    Discuss the specifications that the team decided not to implement and the reasons.
13. **Individual Contributions**
    Describe the major contributions made by each person in the group to the overall effort and the role they played in developing the proof-of-principle (POP) or prototype system.
14. **Financial Considerations**
    Update the financial considerations discussion and multiyear chart (Table 4.15) used in the project proposal that reflects the changes resulting from modifications, corrections, and other adjustments made during the project development phase.

At this point the team is in a better position to assess the project's technical, financial, and economic viability. Time elapsed between the moment the team created the project idea and its completion. Even small shifts in project design and execution may affect the anticipated benefits. If the team engaged in a community-based service-learning effort, describe the benefits realized by the people served.

The discussion may include items that the team finds appropriate such as commercial viability, environmental considerations; sustainability; manufacturability; investment required to produce the product or service; market entry issues; distribution issues; sales potential; return on investment; product or service limitations due to competition, technology, financial viability, or other issues; and reflections on ethical, social, and political concerns.

15. **Summary Feasibility Discussion**
    Describe how the design satisfied or did not satisfy the need identified at the beginning of the effort. Does the team categorize the resulting product, service, or process as a proof-of-principle or prototype and why?
16. **Future Work**
    The end of the project is a better time for the team to assess the project's future opportunities than during the initial proposal phase.
    (a) Should this or another team take the effort to the next step?
    (b) Describe further steps that may be necessary to prove the concept.
    (c) Describe possible future research or development activities.

    One way to identify future work is to perform a Failure Mode and Effects Analysis (FMEA) as discussed in the appendix. The FMEA process is applicable to the design of products, processes, and services. The FMEA is a universal tool that is used in any industry or service, where risk of failure has detrimental effects on the users of a product, process, or service. The primary reason for performing a FMEA is to take action to prevent a failure, improve a design through testing or evaluation, or improve a process through inspection. Healthcare institutions and hospitals use the FMEA concept to make their processes safer for patients. Spotting areas of risk and suggesting methods for improving the design or process not only improves quality but also identifies future work.

    The House of Quality (also discussed in the appendix) helps facilitate group decision-making regarding a product, process, service, or system. It is used to determine development characteristics that combine technical requirements with customer preferences. The collaboration may lead to new ideas for the product, process, service, or system under consideration.
17. **Discussion, Conclusions, Lessons learned, and Recommendations**
    Describe the issues that arose during the development phase and how the team addressed them. Comment on the number and frequency of team meetings, the division of labor, purchasing and reimbursement issues, specification development, testing, report writing, outsourcing issues, mentorship guidance, and course length. Discuss lessons learned from participating in this project. What could be done to improve the capstone course process?
18. **References**
    Provide a bibliography listing all references used for the literature search and background work, the specification of parts, cost comparisons, etc.
19. **Appendices**
    As previously described, appendices may include any supporting material that would detract from the flow of the report if included in the main body.

Final Presentation

After writing the final report, the final presentation and demonstration to the stakeholders should be easy. The final presentation is similar to the proposal presentation, but now the team discusses outcomes and definitive data to support the information previously presented.

Final Presentation
1) Title slide with project team members and date.
2) Background (Optional): Inform the audience about the history or driving force behind the need.
3) Problem Definition and Need: One to three slides showing the problems/needs of current product, service, process, or offerings.
4) Project Description and Objectives: Describe the project.
   a) Include novel features.
   b) List the objectives and the accomplishments.
5) Project Deliverables.
6) Project Scope.
7) Design Overview: One to three slides describing the technical design. Include a block diagram.
8) Test Methodology.
   a) Did the project outcome meet expectations? Explain.
   b) Discuss the results of the test verification matrix.
9) Business analysis: Answer as appropriate. Refine the material presented in the proposal presentation.

During the final proposal presentation, the team may want to update and show a multiyear financial planning chart (Table 4.15) that summarizes the project's anticipated initial costs, revenue, and profit potential. Having worked on the project for several months the team should be able to refine the estimates previously made. As before, some of the line items may not apply to the product, service, or process on which the team is working. Selectively choose the line items, research the costs and sharpen the projected sales estimate. This analysis represents a starting point for the team to determine if this is a financially viable opportunity.

10) Describe the issues and risks that the team overcame.
11) List key milestones completed.
12) Demonstrate product or service.

The written report, specification and test verification matrix should be made available to the stakeholders at the meeting.

Project Development   5

> *Engineering is the art of organizing and directing men and controlling the forces and materials of nature for the benefit of the human race.*
> —Henry G. Stott

> *In mathematics you don't understand things, you just get used to them.*
> —John von Neumann

Chapter Objectives

After studying this chapter, you should be able to:
Understand the phases of the engineering design process.
Understand the cost impact of design errors.
Gain familiarity with alternate engineering design approaches.
Understand the difference between design verification and validation.
Evaluate a design with the use of a test verification matrix.
Implement the engineering design process with your team.

The capstone experience requires students to integrate principles, practices, theories, and methods learned in their academic and work life. Students analyze, synthesize, and evaluate technical and business information in projects that have a professional focus. Students are expected to prepare a proposal that will enable the team to gain approval to move forward with the project. After completing the technical engineering and business effort, they communicate the project results in a final report, presentation, and demonstration.

The Engineering Design Process

To a great extent, the ability of an engineer to design defines the practice of engineering. The engineering accreditation criteria require that a student have a design experience. An engineering undergraduate must have had a major design experience (ABET, Criteria 3 and 5) "based on the knowledge and skills acquired in earlier course work and incorporating appropriate engineering standards and multiple realistic constraints."

Engineering design typically involves the integration of knowledge that considers alternative solutions by selecting the optimal solution with a fixed goal or specification in mind that ultimately results in a product or a service [51]. Design engineering is an iterative, decision making process in which the engineer deals with compromise and optimally applies previously learned material to meet a stated objective. The designer balances product or service flexibility, performance, resource availability, and cost in each step of the engineering design process.

The very act of organizing a design team means that certain design decisions have already been made, explicitly or implicitly. Given any design team organization, there is a class of design alternatives that will not be pursued because the necessary technology, information, and communication paths may not exist. All design teams exhibit preferences based on the participants' backgrounds.

Originally conceived by Conway [52] and subsequently coined Conway's Law by Brooks [53], the law predicts: "Organizations which design systems are constrained to produce systems which are copies of the communication structures of these organizations." Conway goes on to point out that the organization chart will initially reflect the first system design, which is almost surely not the right one. The design will change and the organization must be prepared to change. Conway states that "Because the design that occurs first is almost never the best possible, the prevailing system concept may need to change. Therefore, the flexibility of the organization is important to effective design."

Broaden Conway's law to incorporate the idea that a design will reflect the organization and composition of a capstone project team. Different teams will produce different design approaches and there is not one correct way to complete the effort. Teams must engage in constant re-evaluation and iterative changes as part of their approach to problem solving for large-scale, complex, and sometimes ill-defined systems as well as small and seemingly straightforward engineering applications.

Gassert and Enderle [54] define engineering design as "the creative process of identifying needs and then devising a product to fill those needs." The capstone design project is intended to satisfy ABET Criterion 3 and 5 [4], which expects that students "design a system, component, or process to meet desired needs within realistic constraints such as economic, environmental, social, political, ethical, health and safety, manufacturability, and sustainability." Industry regards a specification (a statement of needs) as a serious commitment—both legally and morally. Failure to meet the specification is regarded as a failure of the design process. The reason for this involves cost. The costs to fix engineering errors increase as the project matures. The results of a NASA report [55], shown in Table **5.1**, illustrates that

Table 5.1 The cost of correcting engineering design errors during project phases

Project phase	System design correction cost factor
Requirements definition	1
Design	3–8×
Prototype fabrication	7–16×
Test	21–78×
Manufacturing	29–1615×

cost factors to correct engineering design errors can increase by orders of magnitude once the effort goes beyond the specification preparation phase. Clearly time should be allotted by the project team to define the problem and prepare an accurate requirements definition to save time and money in the latter parts of the project lifecycle.

As shown in Fig. 5.1, a complete design development has nine phases:
1. Discover a need.
2. Define the problem, assess the capability of the team to carry out a solution, and prepare a comprehensive specification.
3. Investigate alternative solutions.
4. Research that includes a literature review, assessing the competitive landscape and resolving potential intellectual property conflicts.
5. Engineering design.
6. Implementation.
7. Integration, test, and evaluation.
8. Installation (including legacy system data migration for database applications), and
9. Training.

Each phase must be continually reviewed and re-evaluated. However it should be noted that the complete design process is different for different industries, products, customers, and projects.

The NASA Design Approach

The National Aeronautics and Space Administration (NASA) advocates an eight-step engineering design process leading to the development of a new product, service, or system ([56] and [57]). The approach is summarized in the following list:
1. Identification of the problem.
2. Specify the design requirements (criteria).
3. Brainstorm possible solutions.
4. Build on the solutions coming out of the brainstorm session and begin to narrow the selection.
5. Evaluate the solution possibilities by recording the pros and cons of each design idea.
6. Select an approach.
7. Build a model or prototype.
8. Refine the design.

Fig. 5.1 Complete design process

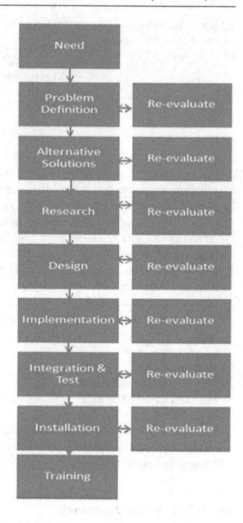

The NASA approach demands that the team clearly identify a problem (step 1) and develop a rudimentary specification (step 2). Based on the team's understanding the target problem, they can then begin to brainstorm possible solutions. The NASA approach in steps 4, 5, and 6 encourages team deliberation and prevents the team from choosing the first likely solution, rather than adequately investigating the situation and its possibilities.

A fault with this NASA design approach is that it does not explicitly mention product or service design verification and validation. The specification establishes the criteria for project success. The design verification process assesses the performance of the product, service, or system against the business and technical criteria or specification.

The Engineering Design Process

Table 5.2 Comparison between design verification and validation

	Requirements	Product or service confirmation
Verification	Do the requirements describe what the team wants to deliver with respect to size, weight, speed, throughput, interface capability, cost, built-in-test, safety, environmental requirements, reliability, software selection, maintainability, output power, torque delivered, etc.?	Is there objective evidence that the product or service satisfies the requirements?
Validation	Are the requirements the right requirements, i.e., do they properly represent the customer's needs and expectations?	Does the product or service, when operated by users in the operational environment, satisfy the customer needs?

Design Verification and Validation

Design verification answers the question "Did we build the System Right?" Each specification requirement must be verified by test, analysis, inspection, demonstration, or similarity. The rule is "test wherever possible." Perform analysis or inspection, where testing is not desirable or possible.

Validation is an examination of the overall system whereas verification is typically an examination of specific requirements. Design validation answers the question—"Did we design or build the Right System that will satisfy the user/customer/client's needs?" Design validation shows that the design meets the requirements **and** the customer's goals and can be produced within targeted cost, schedule, and risk constraints. Product or service validation methods include analysis, simulations, mathematical predictions, trade studies, and tests that examine the specification, design, documentation, and behavior of both the hardware and software. Table **5.2** summarizes the difference between verification and validation.

Sometimes the word robust is associated with validation. Very often a customer requests that a design be "robust" – that is the design should be able to withstand a number of hardware and software failures as well as operator errors and still function as intended. This is a qualitative term and may not always be able to be measured. The word robust, when used with regard to computer software, refers to software that performs well not only under ordinary conditions but also under unusual conditions that stress its designers' assumptions. Confirmation of the design's ability to survive is frequently done by intentionally inserting hardware and/or software faults and observing the system's operation. Reliability, availability, and maintainability may also be part of a validation phase. System dependability involves identifying potential faults and then proposing methods that reduces the risk of the hazard occurring.

Although important, validation is not typically done in a capstone project.

Design Verification Plan

Design verification is an essential step in the development of any product, service, or process in industry. Design verification ensures that the finished item does what it was intended to do. A design project not completely tested may result in an item

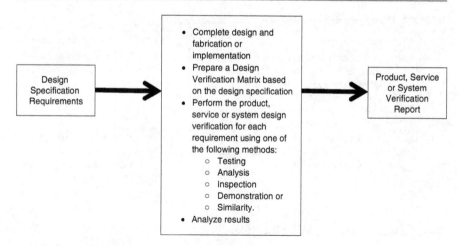

Fig. 5.2 Product, service, or system verification process

that does not meet the customer's specification. Problems discovered after delivery may require costly design modifications. The product design verification process is shown in Fig. 5.2.

There should be a design verification plan associated with the capstone project and it should be done at the end of the effort but before the final presentation and project report are written. The design verification plan establishes with objective evidence that the product, service, or system performance conforms to the agreed upon design requirement or specification.

Product or service design verification is typically accomplished with the use of a Design Verification Matrix (Table **5.3**). That is, each requirement in the specification is listed and the method of verifying that the design meets the requirement is identified in the appropriate column.

There are a number of methods that can be used in verification compliance. Some are relatively inexpensive and quick, such as inspection, while others can be costly and quite involved, such as functional testing. The five common verification methods are testing, analysis, inspection, demonstration, and similarity.

Testing

Depending on test complexity as well as the availability of equipment and a facility, testing can be expensive. However, it is a way to definitively verify that the design meets the specification.

Testing may use special equipment or instrumentation for the evaluation of the product, service, or system to determine compliance with requirements. The test may operate all or part of the unit under test under a limited set of controlled conditions to determine that quantitative design or performance requirements have been met.

For example, a product may be required to operate within a temperature range from 32 to 100 °F. The team could obtain an environmental test chamber that

Table 5.3 Example product design verification matrix

Product or system name:		Test date:					
Test team members:							
Requirement no. or specification paragraph no.	Brief requirement description	Brief verification method description				Pass/fail	Comment
		Test	Analysis	Inspection	Demonstration	Similarity		
1	Size Height = 6″ Length = 8″ Width = 12″			Measured the unit dimensions			Pass	Within tolerance of size requirements
2	Weight—2.5 pounds			Measured the unit weight			Pass	Within tolerance of weight requirements
3	Device is waterproof when submersed in 2 ft of water for 5 min	Place unit under test into a container that can hold 2 ft of water and let it remain in container for 5 min					Pass	Unit under test successfully operated after extraction from two feet of water
.	.						.	
.	.						.	
.	.						.	

controls humidity and temperature. This specialized equipment could be rented but would be expensive. However, for the purposes of a capstone "proof-of-principle" or POP project, the team might consider using a household freezer and a cooking oven with an accurate thermometer to perform the test.

As another example, the project team may have developed an ankle/foot prosthetic. Testing for this project would involve the application of precise forces and motions in line with ISO 22675 and ISO 10328 procedures and FDA guidelines. The project requires test equipment that would deliver rocking motions as well as static load tests for the toe and heel in order to accurately duplicate the field performance of the prosthetic and ensure that the product met its specification objectives. The cost to acquire or rent this test equipment would likely be prohibitive for a college capstone project. The team would have to develop a work around test method.

As a final test example, suppose the team has developed a new barcode scanner. Barcode swipe speed and distance from the device would likely be incorporated into the specification. The tests could proceed by evaluating (1) the reading of barcodes imprinted on different materials, (2) reading the barcode at scanning distances ranges as defined in the specification, and (3) reading the barcode with different swipe speeds from different read ranges. Equipment would be obtained to conduct these tests.

Analysis

Analysis is the use of technical or mathematical models or simulations, algorithms, or other engineering procedures to confirm that the product, service, or system meets a requirement.

Analysis may be used to support the requirement that a product will have a mean time before failure of 10,000 h. A reliability analysis using data representative of the failure rates for various components. Sources of failure rate data to perform reliability verification calculations may be found in the following manuals:

- RIAC Reliability Toolkit—Commercial Practices Edition. Recent data books can be found on the Reliability Information Analysis Center (RIAC) website (http://www.theriac.org/), which is the Department of Defense (DoD) chartered Center of Excellence in the fields of reliability, maintainability, quality, supportability, and interoperability.
- Guidelines For Process Equipment Reliability Data With Data Tables, 1989 American Institute of Chemical Engineers, Published Online: 28 Sep 2010.
- IEEE standard 1413–2010—IEEE Standard Framework for Reliability Prediction of Hardware IEEE, New York, NY.
- IEEE standard 493–2007—IEEE Recommended Practice for the Design of Reliable Industrial and Commercial Power Systems.
- IEEE standard 577–2012—IEEE Standard Requirements for Reliability Analysis in the Design and Operation of Safety Systems for Nuclear Power Generating Stations.

An analysis frequently requires the collection of a variety of data, which individually does not provide a pass or fail indication. The results of all of the information taken together determine whether the item satisfies the success criteria.

Inspection

Inspection involves verification by observing that the requirement is met using one or more of the five senses, simple physical manipulation, or mechanical or electrical measurement. Inspections include a visual check or review of documentation such as, drawings, vendor specifications, software version descriptions, computer program code, data, or the product or service without the use of special laboratory procedures or equipment. Inspection includes examining a physical attribute such as dimensions, weight, physical characteristics, color, or markings.

Demonstration

Demonstration is the actual operation of the product or system to provide evidence that it accomplishes the required functions under specific scenarios. Demonstration typically confirms requirements through the operation of the process or unit under test. Demonstration verifies system characteristics such as human engineering features, throughput, visible features, and transportability. Demonstration relies on observing and recording a product or service operation. It should not require the use of elaborate instrumentation, special test equipment, or the analysis of a great deal of data.

As an example, if a specification for a product requires that it operate with a USB device, then confirm the device is operable after it is plugged into the system. If a device is to be operable with one hand, then someone should operate the product with one hand.

Similarity

If a design includes features or materials that are similar to those used in another product that has met or exceeded the specifications, an analysis to illustrate this similarity may be used to verify a requirement. For example, if a specification requires that a product be water resistant and materials that have been proven to be water resistant in other applications have been chosen, a statement of similarity to the first design could be used to prove conformance to the required specification.

DRIDS-V Design Approach and Plan

Define, Research, Identify, Decide, Solve, and Verify—DRIDS-V [58]—is a six-step design approach that incorporates the ideas presented in the NASA plan but adds the all-important design verification activity.

DEFINE

Identifying a project topic is difficult. The team will brainstorm to come up with ideas. It will leverage its collective thinking by engaging with each other, listening, and building on ideas. The team will conceptualize a product or service that fulfills a commercial, societal, or personal need. Don't discount the development of low-tech ecologically or environmentally friendly products or services. The team must carefully define the problem, question, or goal that it will confront. The team will identify constraints and evaluation criteria. In the background of its thinking, the team must take into consideration the availability of resources required to implement the project. The project should be a challenging task that is achievable within the allotted time frame and within the budget constraints.

RESEARCH

Learn about the target subject. Gather information. Be sure there is a need for the product, service, process change, or system under consideration. What has been done by other academic and industry researchers? Understand the capabilities of similar commercial products and services. Identify patents that may be similar to the team's problem solution. List the team's assumptions.

The market's needs, target pricing, promotional mix (telephone and personal sales calls; advertising using Internet ads, direct mail or e-mail letters, TV and radio ads, newspapers; trade shows; posters and flyers), and distribution of the resulting goods or services must be considered for the project to be viable. Solicit feedback from prospective customers and understand how the target product or service meets their needs.

The selected project should be different in some way from work that has preceded it. It should have a well-defined objective and solve a "real-world" problem. Engineers frequently enjoy the challenge of making a state of the art project. Answer the question: Do all customers want the most sophisticated, highest quality products and services? The answer to this question may influence the design selection and approach. Sometimes the customer wants a basic low cost functional unit. Engineers love to play and provide the customer with a unit that has a host of "bells and whistles" but more often than not, the customer wants a quality product or service that is functional, reliable, and inexpensive.

IDENTIFY

Brainstorm and identify alternative solutions. Compare the strengths and weaknesses of the alternative solutions. How much calendar time will each alternative take? In the team's deliberations make certain that it considers the risks associated with each alternative solution. Develop broad conceptual mitigation plans to overcome the risks for the alternative designs.

List the resources required for each solution. Does the team have the knowledge required to complete the project? Does the team have a sufficient number of skilled people (labor) to complete the required tasks within the allotted time frame? Can a member of the team rapidly learn a new competency, if needed? So much of what teams need to know is not obvious at the beginning of a project. Team members will use the capstone course as an illustration of a lifelong continuous learning process. The course should be an aid to the development of problem solving abilities. The team must confirm that it will have access to needed space, equipment, software, development environments, and materials—when it needs them?

Don't go into analysis paralysis seeking the perfect solution to a problem. Remember there are no perfect solutions. There may be solutions that work better than others do but there is not one perfect solution. If the solution appears reasonable to the team, don't worry about what others will say. Reach a balance between thinking and taking action. So if an approach appears to meet budget, schedule, and technical requirements and can be completed with the available resources, then go for it.

DECIDE

Pinpoint the most important criteria used in the identification of the problem and use those to select the project. Understand what the team will do and what it will not do. Agree on the deliverables. Make sure the team limits the scope of the problem so that it can be completed within the allotted time frame. Once the team agrees on the project, prepare a detailed project schedule listing all the tasks and activities required to complete the project, as well as the relevant milestones, dependencies, and needed resources.

The team will write a project proposal that will be submitted to the mentor for approval. It will be similar to the outline discussed in detail in Chap. 4 and briefly shown below:
1. Brief statement of product or service.
2. Statement of need.
3. Objectives.
4. Technical description (Including a specification).
5. Management.
    (a) Team Members.
    (b) Present a brief team or organizational background that to show that the team possesses the qualifications for completing the project.
    (c) Quality—standards used and a discussion of the type of tests that will be performed to verify that the specification will be met.
6. Resources required—money, people, equipment, software, tools needed to complete the project.
7. Schedule.
8. Budget.

SOLVE

The implementation process follows the project management model of:
- Define.
- Design.
- Do—carry out the required activities.
- Deliver—test and ensure the project meets the expected outcomes.

However, efforts to solve the problem should not move forward until a set of requirements is prepared. Prepare a specification for the proposed project and complete a list of tasks or work breakdown structure (WBS). In its deliberations, the team must be mindful of the interaction between Alice and the Cheshire Cat:

> Would you tell me, please, which way I ought to go from here?
> 'That depends a good deal on where you want to get to,' said the Cat.
> 'I don't much care where—' said Alice.
> 'Then it doesn't matter which way you go,' said the Cat.
> '—so long as I get SOMEWHERE,' Alice added as an explanation.
> 'Oh, you're sure to do that,' said the Cat, 'if you only walk long enough.'
> Alice's Adventures in Wonderland, by Lewis Carroll

The team must define and control its own direction and create a specification that will enable it to move swiftly forward toward a desired destination. Design to the specification!

Solve the problem using the skills, knowledge, and abilities that the team members have gained from courses taken in school as well as the experiences in their work history. Complete the work and document the solution.

The engineering design process rarely moves in a linear fashion. Instead designers iterate, that is they jump back and forth between the steps as they move toward an optimal solution. Designers continually juggle the available resources and technology towards an acceptable end. Ultimately the systems, services, processes, or devices will perform the desired functions and meet a set of needs.

VERIFY

The increasing complexity of today's designs has only served to further intensify the need for a functional verification. Any strategy for product, service, or system success must include the successful completion of a design verification test matrix, which would help to ensure that the solution solves the problem.

The Verification Test Plan provides guidance for the team and the faculty mentor. It establishes a plan to communicate the nature and extent of testing necessary for a thorough evaluation of the product, service, or system. Passing the design verification matrix (Table **5.3**), which evaluates each specification item, proves that the deliverable product, service, or system accomplishes what the team set out to do.

Innovative Capstone Project Examples 6

> *The ideal engineer is a composite. ... He is not a scientist, he is not a mathematician, he is not a sociologist or a writer; but he may use the knowledge and techniques of any or all of these disciplines in solving problems.*
> —Nathan Washington Dougherty

> *Intellectuals solve problems, geniuses prevent them.*
> —Albert Einstein

Chapter Objectives

After studying this chapter, you should be able to:
Understand the components of the product, process, or system lifecycle.
Gain an understanding of common capstone team issues.
Gain ideas for a team capstone project.
Understand the elements of a successful project.

Product or System Lifecycle

Completing a capstone or senior design project is the culminating experience for students in an engineering program. Students apply their knowledge to a team project and the team works together to experience a realistic engineering problem involving design, team work, project execution, and management. Most projects will have common components such as problem definition, marketing, research, scheduling, solution analysis, vendor involvement, design, and the communication of results. The student will become involved in the first eight components (Fig. 6.1) of a product or system lifecycle:

1. Identify a need.
2. Perform a requirements analysis.

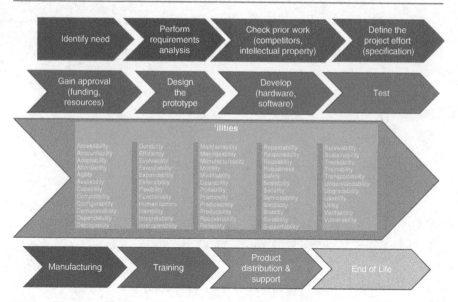

Fig. 6.1 Product or system lifecycle

3. Check prior work (competitors, patents, etc.).
4. Define the effort (specification).
5. Gain approval (funding, resources, etc.).
6. Design the prototype.
7. Develop (hardware, software)—ease of maintenance, durability, reliability, product safety, convenience in use (human factors engineering), aesthetic appeal, and economy of operation.
8. Test.
9. 'Ilities.
10. Manufacturing.
11. Customer product, process, or system training.
12. Product distribution.
13. End of life.

Students typically seek projects involving the development of a proof of principle where the business thinking (business plans and marketing strategies) are at an early stage. The team may attempt a project innovation that will help move the concept to the point of commercial launch, sale, license, or presentation to a venture capitalist.

Students focus on the first eight topics in the product or system lifecycle (Fig. 6.1) during the capstone course but the engineer's job in industry is not over after the prototype is complete. The next steps are as important as the design and prototype effort. Suppose an industry project team completes a product or system and delivers it on time, within budget, and correctly performing all its specified operations. Is it a good product or system? Will the customer be satisfied? Are there other product or system attributes that should be considered? Is the product easy to use - or misuse? Does it require constant maintenance? Is it robust, reliable, and secure? Is it difficult to integrate with other hardware or software products? The practicing engineer must

Table 6.1 Common 'ilities attributes

Attribute	Description
Adaptability	Ease of introducing new features into the product, process, or system
Availability	Percentage of time that the product, process, or system is operating versus nonoperating
Durability	The ability of a product, process, or system to survive its intended use for a suitable long period of time while providing the intended levels of maintenance and operating in the intended conditions of use
Human factors	Human usability. Designing products, processes, and systems that fit the human body and its cognitive abilities. Frequently involves ergonomics issues
Interoperability	Corresponds to the effort required to link to other software, hardware, products, processes, and systems
Maintainability	The totality of design factors that allows maintenance to be accomplished easily Preventive Maintenance—the time and effort required to reduce the risk that the product, process, or system breaks down Corrective Maintenance—the time and effort required to locate and fix an error such that the product, process, or system is repaired and restored to its fully functioning state
Operability	Ease of operation
Reliability	Performance of the product, process, or system to function without failure in a defined environment for a designated period of time
Resilience	The degree to which the product, process, or system can recover quickly from a major disruption
Robustness	The ability of a product, process, or system to operation as intended even when conditions change such as the application of invalid inputs
Safety	Reducing the opportunities for human error in the design and use of products, processes, and systems
Scalability	The property of reducing or increasing the scope of processes and capability according to the product, process, or system needs
Security	Detection and protection mechanisms that protect the integrity of products, processes, and systems
Supportability	A broad term involving post-production operational issues focusing on minimizing life-cycle costs and maintaining the operational readiness of the product, process, or system
Sustainability	The integration of economic, environmental, and social elements so as to minimize product, process, or system waste in design, construction, and use
Testability	The effort required to test a the product, process, or system to ensure correctness
Traceability	Ability to trace the product, process, or system components back to the requirements
Understandability	Ease of understanding the function of a product, process, or system
Usability	Effort required to learn and operate the product, process, or system

keep in mind the remainder of the product life cycle effort – especially the 'ilities. The design is not done until the 'ilities are done.

The ilities attributes usually but not always end in "ility." Attributes such as **maintainability, reliability, interoperability, and scalability** are properties of products, processes, and systems that may not be part of the primary functional performance requirements—but they are often major cost drivers and they may be very important to the customer. Ilities frequently impact product and system design as much as the primary functional requirements. The ilities attributes sometimes represent a challenge to quantify and measure. Human factors engineering may be required to assess the attributes. Some of the common ilities attributes are described in Table 6.1.

Common Capstone Team Issues

Capstone team members have to be wary of a variety of issues that arise and delay project progress. These issues will vary depending on the type of capstone course—for example, competition-based project teams, projects built upon the work of prior teams, and project teams permitting students to join in the midst of the effort. At different colleges, capstone courses differ with respect to available resources, faculty involvement, project cost support, student backgrounds, sponsor involvement, and industry support and participation. A typical capstone course experience involves a team of students working on a design project for two semesters. Common issues among project teams frequently fit into one of the following ten areas.

1. A lack of knowledge about the business or industry into which the product, process, service, or system exists. Investigate intellectual property that relates to the project. Investigate competitive products to understand similarities and differences with the team's potential product. Understand where your product, process, service, or system fits.
2. Starting a design before knowing what the team really wants to do. Develop as complete a specification as possible before designing.
3. Design Decisions—Conflicts or problems arising from an inability to agree on project-related issues including scope, criteria, and alternative approaches [59].
4. Group dynamics difficulties involving different skill sets of people on the projects and having difficulties integrating them into a useful functioning team [60].
5. Building team consensus.
6. Not preparing a risk mitigation plan for some of the most obvious potential problems.
7. Workload imbalance—Conflicts arising as a result of a team member who takes on too much or too little responsibility [59].
8. Capability deficiency—A team member who cannot effectively do the assigned work because he/she does not have the requisite competencies or skills [59].
9. Miscommunication—Conflicts that arise from misunderstandings, failure to provide information to one another, or failure to understand one another. Making changes without informing the remainder of the team [59].
10. Personality—Team members who cannot get along [59].

Be aware of the foregoing danger signs. The team project effort will be successful if people can put personalities aside, cooperate with each other, capitalize on one another's technical and nontechnical strengths, and forge a strong crew based on mutual respect and trust.

Capstone Project Examples

Selecting capstone project topics is difficult. Consider using the methods of the House of Quality and the Failure Mode and Effects Analysis described in the appendix to select a project topic. It may be useful as well as satisfying to investigate and select service learning projects that will help people in your town or elsewhere. Students already working in business and industry should consider proposing job-related research and development topics that would not only be a practical effort but might also serve to improve their status at work. Contact local companies to determine if there exists an industrial problem that is relevant to a company's business needs. A company may wish to collaborate with the team and provide funding, equipment, and technical engineering assistance. Be sure to evaluate the suitability and scope of projects obtained from industry. Keep in mind that the ownership of the intellectual property resulting from the solution to the problem should be discussed and resolved before the industry collaborators begin work. It must be made clear to industry partners that projects are primarily for educational purposes and neither the team nor the university is liable if a project does not produce the desired outcomes.

The project's scope will almost certainly change as the team learns more about the topic. Students frequently overestimate their technical abilities, underestimate project complexity, uncover intellectual property that bears on the topic and forces a change, and underestimate the time required to complete various tasks. Sometimes as the project direction turns, the team suddenly discovers they lack certain skills and must compensate for this problem.

A major challenge for students in the capstone course is to maintain a high level of motivation. Over the project time span, there will be peaks and valleys in enthusiasm as forward progress is advanced or not.

The capstone project will be one of the most valuable learning experiences encountered by students in the pursuit of their degree. The practical skills obtained during team projects represent one of the prime benefits of the capstone course. The quote attributed to Confucius (551–479 B.C.E.) sums up the benefits of the experience: "Tell me, and I will forget. Show me, and I may remember. Involve me, and I will understand."

Review the following sample of projects that addresses a host of needs in a variety of disciplines to stimulate the team's thinking. Some of the projects listed below involve web applications and knowledge of web marketing techniques. Many demanded special technical skills such as welding and programming competencies. All the projects required contributions from people having special knowledge, different backgrounds, and skills. Some projects were more successful than others. All of the projects required scope changes as the effort progressed and teams encountered unexpected issues. Flexibility and compromise was demanded. Some of the projects were serviced based and benefited towns and communities. Two projects received extensive employer support and resulted in dramatic cost savings after implementation. Some of the projects were truly innovative and others represented improvements to existing products and services.

Web-Based Game

The team designed and developed a fun and interactive mobile application to encourage children to complete chores assigned by parents, behave and perform well in school, and enable the parent to manage these tasks with a minimum of effort.

Town Intranet Portal

The team developed a user-friendly method of accessing needed data across multiple shared town networks and databases. The team provided the town a vehicle for internal communications whereby all employees could post and reply to messages so that information could flow more freely within town government.

International Student Transition System

Created an information website for the purpose of informing prospective international college students about the culture, language, education, and other differences between their country and the United States.

Hot Water Heater Information Tool

Developed a web-based tool to provide energy contractors, consultants, and professional installers the ability to get a report on the savings, installation, and operational lifecycle cost of implementing different hot water systems.

Biometric Signature Solution for a State Medical Facility

Healthcare facilities have a need for HIPAA compliant electronic biometric signature system to confirm a provider's written diagnosis treatment plan placed on a computer. This team decided to use an electronic fingerprint sensor device placed on each provider's computer. It implemented an electronic bio-metric verification of clinicians who sign patient's progress notes. A psychiatric State Medical Facility that served over 2,400 patients annually was used to test the system. Previously, the physicians, therapists, clinicians, and nurses signed hard copies of each patient's progress report. These hard copies were then filed in binders. The overall process was time-consuming, expensive, wasteful of space, and error prone. This project was implemented and alleviated many of the issues associated with the signature process at the hospital.

Emergency Operations Center (EOC) Design and Implementation

This project was intended to improve the ability of small towns to operate critical services during weather or other emergencies lasting up to one month. Town officials were consulted in an effort to understand their EOC needs (primary or backup) and budgetary constraints. A baseline Emergency Operation Center design was developed and implemented. A town agreed to be used as a test case and the EOC was implemented at a local university.

Parking Solutions

This project team developed a stand-alone optimized parking management solution for small institutions that interface to an existing computer system. It included features for online new and renewal vehicle registration and identified vehicle parking information. The project made information accessible to customers and permitted parking stickers to be mailed to home address. It included customizable features to offer the flexibility to interface with current institution systems. It was intended to be a cost-effective solution for small to medium institutions.

Phone-2-Phone (P-2-P) Translator Application for a Smart Phone

Multi-language communication presents many challenges for organizations and individuals. A Phone-2-Phone (P-2-P) Translator Application was developed that enabled real-time human phone conversation by automatically translating to and from multiple languages using smart phones.

Chimney Sweep

The lack of periodic chimney cleaning is a leading cause of home fires in the United States. Other than a traditional chimney sweep, there are few options for the do-it-yourself homeowner to clean a chimney conveniently by themselves. The team designed, built, and tested a cost-effective, reliable chimney cleaning system that was easy to use, efficient, and required minimum clean-up.

Green Fertilizer

The Green Fertilizer team developed a self-sustaining fertilizer system that used rain water, solar power, and natural ingredients, which were aerated to create a compost tea. The team constructed and tested the system which was capable of pumping the compost tea to fertilize a grass lawn up to 1/2 acre in size. This system was intended to be effective, cheaper, and environmentally friendlier than common chemical-based fertilizers. The system was compliant with public school state laws.

Rainwater Harvesting System Project

The team designed, fabricated, and implemented a rainwater collection system on the university campus. The design harvested rainwater runoff from the Campus Center's fourth floor roof and third floor patio area. Using existing rain leads and piping, the team collected rainwater in a large storage tank. The rainwater was then pumped to the irrigation line and then used to water the lawn and greenery.

The harvested rainwater was also intended to be used to fill up the university watering trucks that are used to water flowers, shrubs, and greenery over the 200-acre campus.

Helmet Integrated Impact Detection System

Sports-related concussions or mild traumatic brain injuries (mTBI) are reported over 300,000 times per year. It is predicted that over half of the concussions received yearly are not reported to medical professionals. These injuries can be compounded by subsequent impacts within a short period of time. Medical professional, coaches, athletes, and young athlete's parents realize the importance of identifying mTBIs as soon as possible to prevent long term damage. Early identification of mTBI would provide invaluable information to help determine whether a player can safely continue to compete or if they should receive medical attention.

The team designed a low cost modular sensor array to measure the impact received by an athlete. The unit was placed in a sports helmet. The design was able to alert the athlete, coach, or parent via a Bluetooth receiver placed in the helmet that the athlete received an impact greater than a predetermined concussion-probable threshold.

Fitness Monitoring Device with Smartphone Application and Conventional Bluetooth

The team designed and developed a fitness monitoring device that integrated three different sensors (heart rate, pulse oximetry, and muscle activity) that were wirelessly linked to a smartphone application (Android) utilizing conventional Bluetooth technology.

Voice-Controlled Wheelchair

Designed and built a microcontroller-based voice-controlled wheel chair for para- and quadriplegics. A speech recognition system controlled forward and rear movement. The hardware and software also permitted the operator to make left and right turns.

Snow Melt Mat

The team designed, fabricated, and tested a prototype for an affordable and disposable chemical-based snow melting mat for private home walkways. The snow melting capability operated for several hours and was competitive priced with electrically powered snow melting mats.

Snow Load Sensor Mat

The team designed and tested a prototype Snow Load Sensor Pad to measure snow load on flat commercial building roofs. A sensor controlled an alarm that alerted building occupants when the snow load on a roof exceeded a preset limit.

Mini-Hydro Potable Water System

The team developed an inexpensive, reliable, and efficient water filtration system using local resources that could produce potable water for underdeveloped areas or disaster afflicted zones. The approach also used a moderate flow river or nearby stream as the water source.

Fire Hydrant Locator

The team developed, analyzed, and demonstrated a wireless detection system to aid fire fighters in locating hidden fire hydrants during fire emergencies. The approach used selected radio-frequency identification (RFID) tags that could operate while buried under six feet of snow. The RFID approach enabled the location of the hydrants under severe conditions where hydrants could be hidden under deep snow, heavy brush, or weed growth.

Firefighter Training Vent Simulator

Ventilation is the process by which by-products of a fire (heat, smoke, and toxic gasses) are removed from a structure so rescue and suppression crews can safely and easily enter a blazing building. Ventilation can be made by natural or manmade openings in the envelope of the building. Firefighters commonly cut a hole in a roof to allow vertical ventilation.

The team designed and fabricated a ventilation simulator prototype that was structurally stable, cost-effective, provided a realistic experience and had a simulated roof that was easily replaced during a training session. The design accommodated a variety of roofing materials and had an adjustable roof pitch. The vent simulator prototype was designed and tested to support three firefighters in full gear.

The team developed a business model and a product that could contribute to reducing the line of duty death rate among US firefighters. The unit was successfully evaluated for training purposes by more than four Connecticut fire departments.

Parts Protection Bag

The team developed a reusable flexible bag to prevent a variety of damages that occur to machined parts and gears. The project developed a piece-part protection container system that protected against environmental damage (e.g., corrosion) and physical damage such as nicks, dents, and scratches caused by human mishandling or shipping. The container was vacuum sealable as well as waterproof to aid in eliminating environmental damages. It was foam lined to prevent physical damage. Unlike similar products currently on the market that are bulky and used valuable space in factories and storage units, the bags stored pieces using as little space as an average size magazine or book.

Medical Reminder System (MRS)

The team developed an automatic medical office notification system targeted at reducing medical office expenses by providing physicians an effective and efficient automatic method to communicate appointment and instruction reminders to patients using e-mail, text messages, and/or voice mail. The system was intended to do the following:
- Maximize treatment program effectiveness.
- Increase medical office efficiency.
- Lower medical office costs.

PALS: Personal Alert Location System

The team developed a simulation for a Personal Alert Location System that had the capability to aid people in need of emergency assistance. The device enabled individuals to summon help via a call center. The system included embedded location tracking; readily available personal medical information for medical providers; the ability to be remotely turned on or off; and offered silent notification when activated.

High Wind Speed Alternate Energy System

The team developed designed and analyzed a system to use turbines at strategic airport locations to generate electricity from the wind energy created by jet aircraft on take-off. A representative airport was used to determine the return on investment (ROI).

HeDS-UP: Motorcycle HeDS-UP (Helmet Display System)

The team developed an approach to the creation of a distraction-free, fully integrated motorcycle information system contained in a motorcycle helmet that would provide audio and visual cues to the rider.

The HeDS-UP project addressed the problem of motorcycle accidents caused by motorcycle riders losing visual contact with the environment ahead when glancing at the motorcycle's instrument panel to obtain vehicle operational information. This problem can lead to vehicular damage, vehicle operational failure, and/or serious injury or death to the rider.

The HeDS-UP team's approach to address these problems was to develop a visual display and aural annunciation prototype system that was integrated into a DOT/Snell certified motorcycle helmet. The visual display provided the helmet wearer with real time vehicle operation status including: vehicle speed, engine RPM, vehicle status, and warnings. The aural annunciation system provided the helmet wearer with real-time rider selectable audible annunciations of vehicle operation and performance.

Automatic Ladder Leveler

The team developed an innovative, stable, and safe after-market automatic ladder leveler that attaches to extension ladders. When the ladder was placed on uneven ground the leveler quickly and automatically leveled itself. The leveler positively locked as both legs come in contact with the ground.

Solar Powered Food Dehydrator

The team developed a solar powered food dehydrator to preserve fruit and vegetables. The team wanted to make a contribution toward creating a healthy and sustainable food supply. The team used indigenous recovered materials and designed the unit with these qualities:
- Dried food quickly.
- Adjustable vent to control drying temperature and airflow.
- Compact and lightweight for portability.
- Pest-proof.

The Elements of a Successful Project

There is no "silver bullet" that guarantees a successful project. The first challenge the team faces involves project selection. The considerations of project selection can seem overwhelming. Idea gathering, screening, prioritizing, resourcing, tracking, measuring, optimizing—that's a lot of activities to manage.

Subsequently, managing the actual project may sometimes seem like boiling the entire ocean at one time. So many things are happening at once. The team confronts challenges associated with task distribution, scheduling, design problems, hardware availability, software interfaces, system differences, database environments, software versions, competitors, intellectual property, patent infringement, prototype failures, test issues, and people personality conflicts. All of these will seem to intersect at the speed of light. Limit the scope of the activities. It is vital to set reasonable

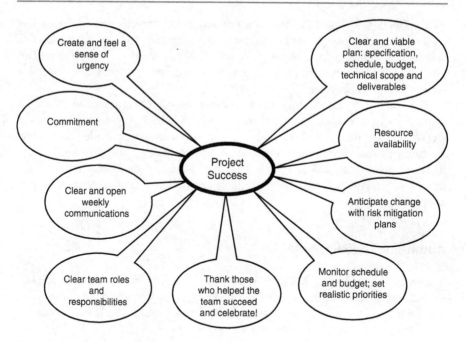

Fig. 6.2 Elements for project success

expectations—what can realistically be done, by when, by whom, and at what cost? But there are steps that the team can take to minimize the risk of failure. There are ways to improve the success of a project (Fig. 6.2). Consider these thoughts.

1. Prepare a clear and viable plan: technical scope, specification, task definition, schedule, budget, and deliverables.

   Adherence to schedules, budgets, and specifications become more effective when they are created collaboratively. If the requirements are unclear, the project should not start. As previously stated, the process of developing a reasonable specification may be painful, but without it the team cannot identify equipment needed, material for purchase, budget, or project tasks. Without a specification and a test plan, the team does not know when to stop working.

2. Resource availability

   Resources include time, money, people, materials, equipment, facilities, technology, information, and space. For information system projects, more specific resources may be needed such as system developers, web developers, database administrators, system analysts, servers, and special development environments.

   Efficient and effective use of resources can make or break a project. Determine the difficulty of obtaining the resources that the team needs. Verify that the team has the budget for them and that their availability fits into the project's schedule.

3. Anticipate and accommodate change.

   The only certainty in life is change—expect it! Change is inevitable and is best when embraced, not fought. Anticipate change with well thought out risk mitigation plans. Identify and evaluate the risk associated with equipment availability, long lead item purchases, weather conditions, vendor issues, schedule and requirements before starting the project.

4. Clear and open communications

   Communicate often – at least once per week. Agile software practices advocate a daily stand-up meeting known as the daily scrum. Teams focus on three key questions: What was done since the last meeting? What will the team accomplish between now and the next meeting? What constraints (if any) impede progress?

   Any news is news (good or bad). Remember Colin Powell's admonition, "Bad news isn't wine. It doesn't improve with age." Be upfront about the project status or the team will not be able to correct it.

   Use a variety of communication tools and know which ones work best for each team member. Long distance communication has become a key to the success of many organizations. Remote workers and virtual teams have become the norm. Emails and other written communications are great for messages without too much complexity, but should not be used simply because it's more convenient or saves team members from having a difficult conversation. The team needs to build trust in one another and emails and remote conversations are not sufficient to do that.

   Misunderstandings and disagreements can occur without meeting in person. Conference calls can lead to misunderstandings either due to lack of communication or simply because the medium is not conducive to individuals asking for better meaning. Body language plays a part in our communication. It is not just what is said; how one says something, facial expressions, and body gestures also bring meaning. This is lost in a phone conversation.

   Face-to-face communication is useful in situations where:
   - People are new to one another. When a team sits in the same room, it encourages people to participate.
   - The team members want to be sure they are on the same page regarding a design approach, or overcoming issues or risks.
   - The team members sense disagreement or conflict, or need to ask for advice or feedback.

5. Monitor schedule and budget; set realistic priorities

   The aim of planning and scheduling is to make certain that the team and all stakeholders understand.
   - The status of the work and where it should stand.
   - The actions that must be taken if delays occur.
   - The cost of correcting delays.
   - The impact of delays.
   - The need to reallocate resources from low priority to high priority tasks in order to minimize the impact of unforeseen issues that may arise.

     Everything can't be critical. Understand the value of each project task and prioritize limited resources.

6. Commitment.

   The team has to support one another and make a commitment to fully support the project. When this occurs, magic can happen!

7. Maintain a sense of urgency in the work.

   In today's business environment, if you don't move fast, you get run over. The same is true in the capstone course. One or two terms may seem like a lot of time but it isn't. Without a sense of urgency, people and businesses can't move fast enough. Speed is the driver because stakeholders frequently have a zero tolerance for waiting, and there are always competitors gaining on you [61]. Grab the opportunities and make something happen today. Restructure plans or even shed low priority activities to move faster and smarter.

8. Clear team roles and responsibilities.

   Distribute the work based on an understanding of the tasks ahead, as well as the preferences, strengths, skills, experience, and interests of the team members. It is possible that some roles and tasks may overlap. With a small group one person may take on more than one role. Define clear responsibilities. If responsibilities are left open and not designated to specific team members, it may result in one or two members doing most of the work. This will create resentment. Make sure that each team member has a strong idea of what his or her responsibilities include. Members should also avoid taking on others' responsibilities without notifying the affected person. This might cause frustration or confusion.

9. Thank those who helped the team succeed and celebrate!

   At the end of the project reflect and take pride in the team's accomplishments. Acknowledge, recognize, praise, and thank all those that played a part in the project's success. Celebrate the *completion of the project. The team deserves it*!

   This last item is often overlooked in industry and academe. Since projects are finite, they *will* have an end...no matter how far off that might seem at the beginning. When the team arrives at this point, make sure that the team identifies some way to celebrate the accomplishment. A celebration doesn't have to cost a lot of money, but recognizing others' contributions and the completion of the goal is important. Jack Welch, one of the foremost industrial leaders of the twentieth century, concurs about celebrating success. He says:

   > Whatever the reason, too many leaders don't celebrate enough. To be clear here, we do not define celebrating as conducting one of those stilted little company-orchestrated events that everyone hates, in which the whole team is marched out to a local restaurant for an evening of forced merriment when they'd rather be home. We're talking about sending a team to Disney World with their families, or giving each team member tickets to a show or a movie, or handing each member of the team a new iPod. What a lost opportunity. Celebrating makes people feel like winners and creates an atmosphere of recognition and positive energy. Imagine a team winning the World Series without champagne spraying everywhere. You can't! And yet companies win all the time and let it go without so much as a high-five. Work is too much a part of life not to recognize the moments of achievement. Grab as many as you can. Make a big deal out of them. That's part of a leader's job too—the fun part [62].

   The college might not send the team to Disney World, but a pizza party seems like something within the realm of possibility.

Intellectual Property 7

> *Engineering is a great profession. There is the fascination of watching a figment of the imagination emerge through the aid of science to a plan on paper. Then it moves to realization in stone or metal or energy. Then it brings homes to men or women. Then it elevates the standard of living and adds to the comforts of life. This is the engineer's high privilege.*
> —Herbert Clark Hoover,
> Engineer & 31st President of the United States

> *Engineering is the professional art of applying science to the optimum conversion of natural resources to the benefit of man.*
> —Ralph J. Smith

Chapter Objectives

After studying this chapter, you should be able to:
Understand the meaning of intellectual property.
Understand what a patent is and how long it is viable.
Understand the reasons to do a patent search.
Learn how to do a patent search.
Describe the types of patents.
Understand the invention protection that an individual can obtain without a patent attorney.
Understand when to obtain professional assistance.
Understand what a patent attorney does.

Intellectual Property

Intellectual property is a term for a creation of the mind. It may be an invention, literary or artistic work. Copyright, patent, or trademark protection gives the creator legal control over when, where, and how his or her creation is published or used in the United States.

A patent provides legal protection for inventions. Title 35 of the United States Code relates to patent law. The law (35 U.S.C. 101) permits patents to be granted only for "any new and useful process, machine, manufacture, or composition of matter, or any new and useful improvement thereof." An invention may be able to be patented if it is novel and nonobvious. A U.S. patent is granted for an invention for a term of 20 years from the date that a person or company files a first U.S. nonprovisional application.

Copyright protection covers material such as novels, scientific papers, books, articles, poems and plays, films, musical works; artistic works such as drawings, paintings, photographs, and sculptures; and architectural designs. In the United States, protection extends for the life of the copyright-holder plus 70 years.

A trademark is protection for a distinctive word, phrase, logo, or other graphic symbol that may distinguish one product, company, or organization from another.

Trade secrets represent confidential information about a process, method, or design that an organization uses and that provides a competitive advantage. Trade secrets are protected by maintaining the confidential information as a secret.

Some entrepreneurs and companies rely on non-disclosure agreements (NDAs) to protect proprietary information. The main point of a non-disclosure agreement is to prohibit improper use and disclosure of information that may be disclosed during discussions about a product. These agreements require the signatories to maintain the confidentiality of specific information for a defined or indefinite term. The agreements are binding only on the parties who actually execute them. NDAs concern only information that is not already publicly available. The protected information should have technical value such as the Coca-Cola Company's secret syrup formula or commercial value such as business plans or financial records.

Patent law requires a public disclosure of the best mode or the best way to make or use the invention. Consequently, people and companies may learn much about the patented invention, but they may not legally use it without gaining permission. Trade secrets are not disclosed to the public. Some companies purposely choose not to file a patent about an invention in order to keep the knowledge secret. An unauthorized disclosure of a trade secret by a person who is obligated to maintain the trade secret is known as a misappropriation of the trade secret. It is therefore unethical, and illegal, for an employee to disclose trade secrets that belong to their employers—even after they are no longer employed by the employer. However, it is not illegal to disclose knowledge that has been patented, because that knowledge is in the public domain. It is also not illegal, although it may be considered unethical, for an employee to disclose or use information that an employer has not designated as secret, confidential, or proprietary.

Intellectual property is big business. Intellectual property rights, like any other property, can be sold or given away or licensed. Public and private companies such as Intellectual Ventures, RPX Corporation, and Acacia Research Corporation acquire, develop, license, manage, and protect intellectual property from unauthorized use. A single alleged infringed claim can cost millions of dollars in litigation and settlement awards. However the litigation can take a long time and be very onerous. As an example, Bob Kearns had been granted a patent for a design of intermittent windshield wipers in 1967, two years before Ford introduced them in production. Ford and Chrysler claimed that they developed their own system and had no need to license Kearns' design. Litigation ensued and it took until 1995 for the U.S. Supreme Court to rule against Ford and Chrysler and award Kearns $30 million. Businesses buy and sell intellectual property regularly. In 2012, AOL agreed to sell 800 patents to Microsoft for $1.1 billion. In 2013, Kodak agreed to sell Apple and Google intellectual property for $525 million.

Patent Search

Why do a patent search in the capstone course? First and foremost, the team does not want to infringe on another company's or inventor's patent rights for legal and ethical reasons. So the team should determine if the project that it is developing has been patented. If the idea has been patented, then the team should either modify its work, or seek a license to the patent, to avoid possible infringement.

Prior art includes previously issued patents, published patent applications, trade journal articles, publications (including data books and catalogs), public discussions, trade shows, or public use or sales of the proposed product or service. In a case where a team is considering seeking a patent for an invention, a thorough search for prior art will help the team members assess the novelty of the invention.

Even if a product does not appear to be on the market, it does not necessarily mean that a patent does not exist. An original creative invention may still have been patented years before the team came up with its inventions. A product may have been patented but never have made it to the marketplace. Thus, a company or individual could unknowingly infringe an existing patent. Placing a new product on the market without first checking its patent status places the company at some risk. Many patent owners are very protective of their intellectual property and will sue a person or company that they believe has infringed their patent. Patent searching at an early stage can save a lot of time, effort, and money.

Some companies have a policy of intentionally not doing patent searches, and try to generally remain unaware of any patents that they may infringe. This is because a company or a person is not liable for patent infringement damages until they are made aware of any patents that they may potentially infringe.

If the results from the patent search show that prior art exists, the team may be able to improve upon the patent. If the refinement is a novel and nonobvious change or improvement, it may meet the standard for a new patent. The team can certainly proceed with the project based on the modified invention.

Doing a patent search will help a team gain confidence in the quality of the invention as the team members become more informed about the current state of the art. If the team goes forward with a patent application, the knowledge gained from the search may make the patent application stronger and less likely to be rejected by the USPTO.

A common way of conducting a search is to do a subject search or a company/inventor search on a patent office website. The United States Patent and Trademark Office (USPTO) website (http://patft.uspto.gov/) can assist the team in obtaining information about inventions that have been patented. Additional information to improve a person's understanding about patents can be found at http://www.uspto.gov/inventors/patents.jsp.

Google also has a patent locator. A YouTube Google patent locator tutorial video may be found at http://www.youtube.com/watch?v=T20l8v40gzY. The Google Patent Search website permits people to search through patents from the USPTO and is located at www.google.com/patents.

Example Patent Search

The Discovery Education 3M Young Scientist Challenge (a science competition for students in grades 5–8) honored Peyton Robertson as America's 2013 Top Young Scientist for developing a novel sandbag. The 11-year-old boy from Florida designed a sandbag to better protect life and property from the ravages of saltwater floods. Instead of sand, his bag was filled with a mixture of salt and an expandable polymer. When dry, it was lightweight, easy to move, and easy to store. In use, the bag would be positioned to create a barrier around a property. The expectation was that the polymer would absorb the approaching water, swell, and fill the volume of the bag. With a sufficient number of bags the property could be protected.

In order to determine if a patent existed for this invention, go to http://patft.uspto.gov/ (Fig. 7.1). Click on Advanced Search. In the Query box, enter the terms sandbag AND salt AND polymer using the logic function AND to denote that a database search should include the three words sandbag, salt, and polymer (Fig. 7.2).

Clicking the search box on the website page results (Fig. 7.3) in a list of 15 patents containing all the words sandbag, salt, and polymer. In order to confirm the existence of prior art, the investigator would review each of the listed patents. In the list of 15 patents in Fig. 7.3, the sixth patent is U.S. Patent No. 7,258,904, a portion of which is presented in Fig. 7.4. The patent abstract in Fig. 7.4 describes the patented invention as follows:

> According to the present invention, a water absorbent material and materials to be sucked included in a water permeable bag type member form a water absorber. Thus, the water absorber is small both in weight and volume upon its transportation, and accordingly, the water absorber can be rapidly conveyed in large quantities without depending on a human power. The water absorber absorbs water upon its use to adequately satisfy a function of a form traceability for weight, volume and outline.

Patent Search

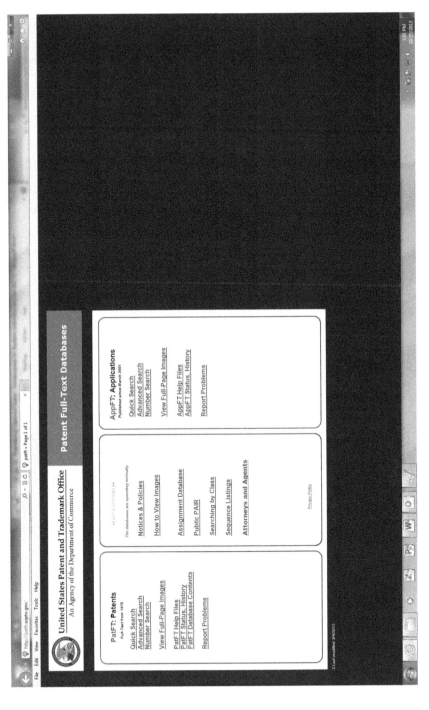

Fig. 7.1 The United States Patent and Trademark Office website

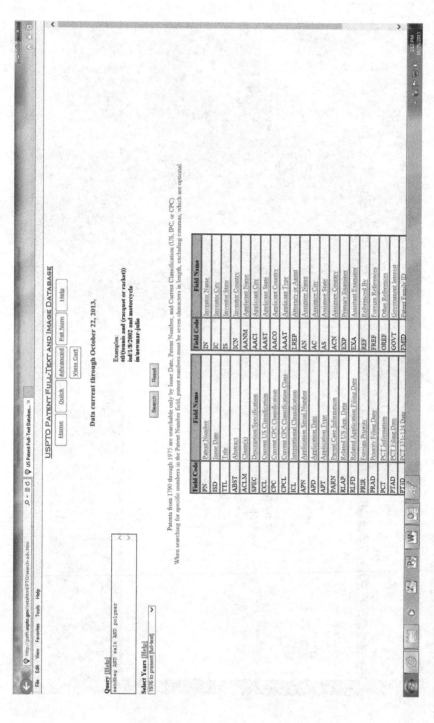

Fig. 7.2 Searching the USPTO database

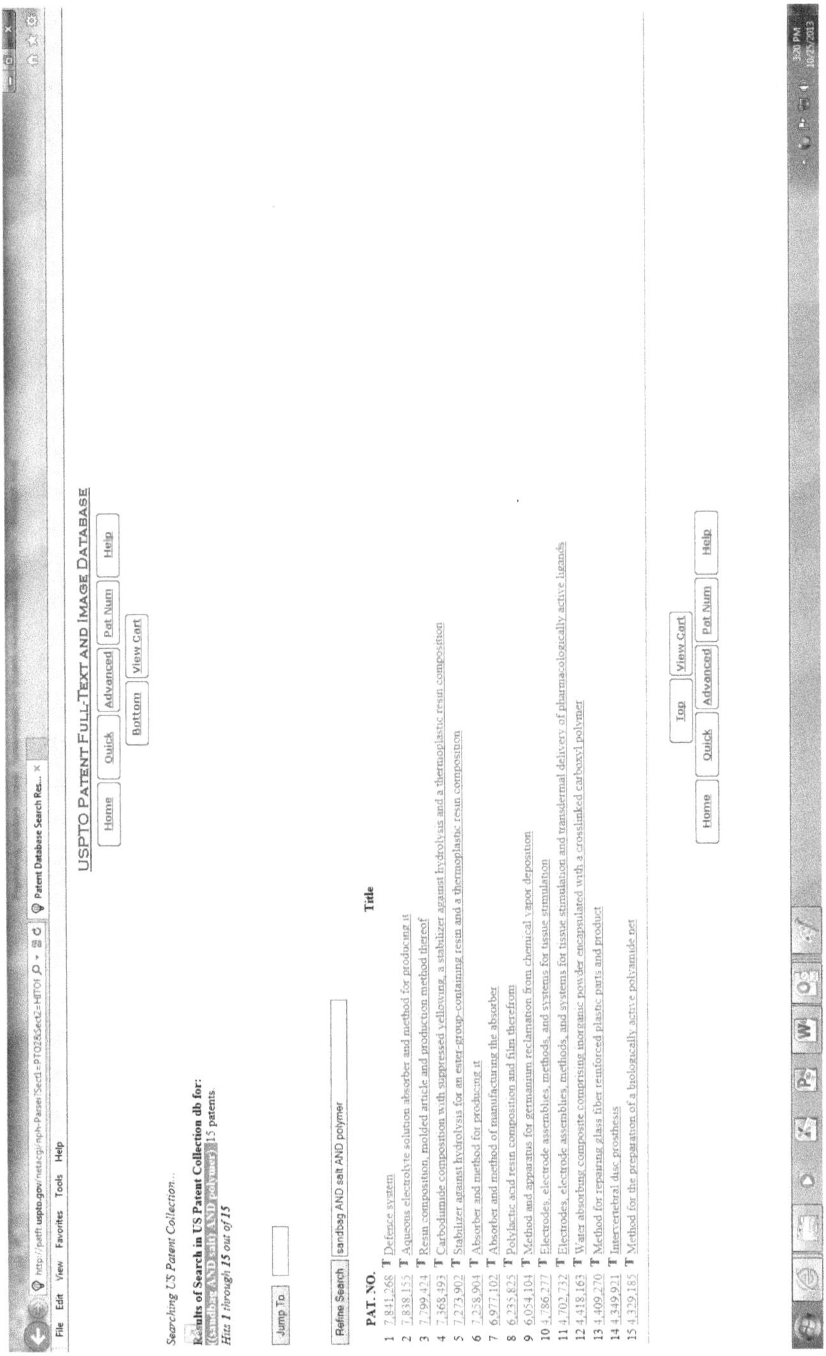

Fig. 7.3 A list of patents containing the words sandbag, salt, polymer

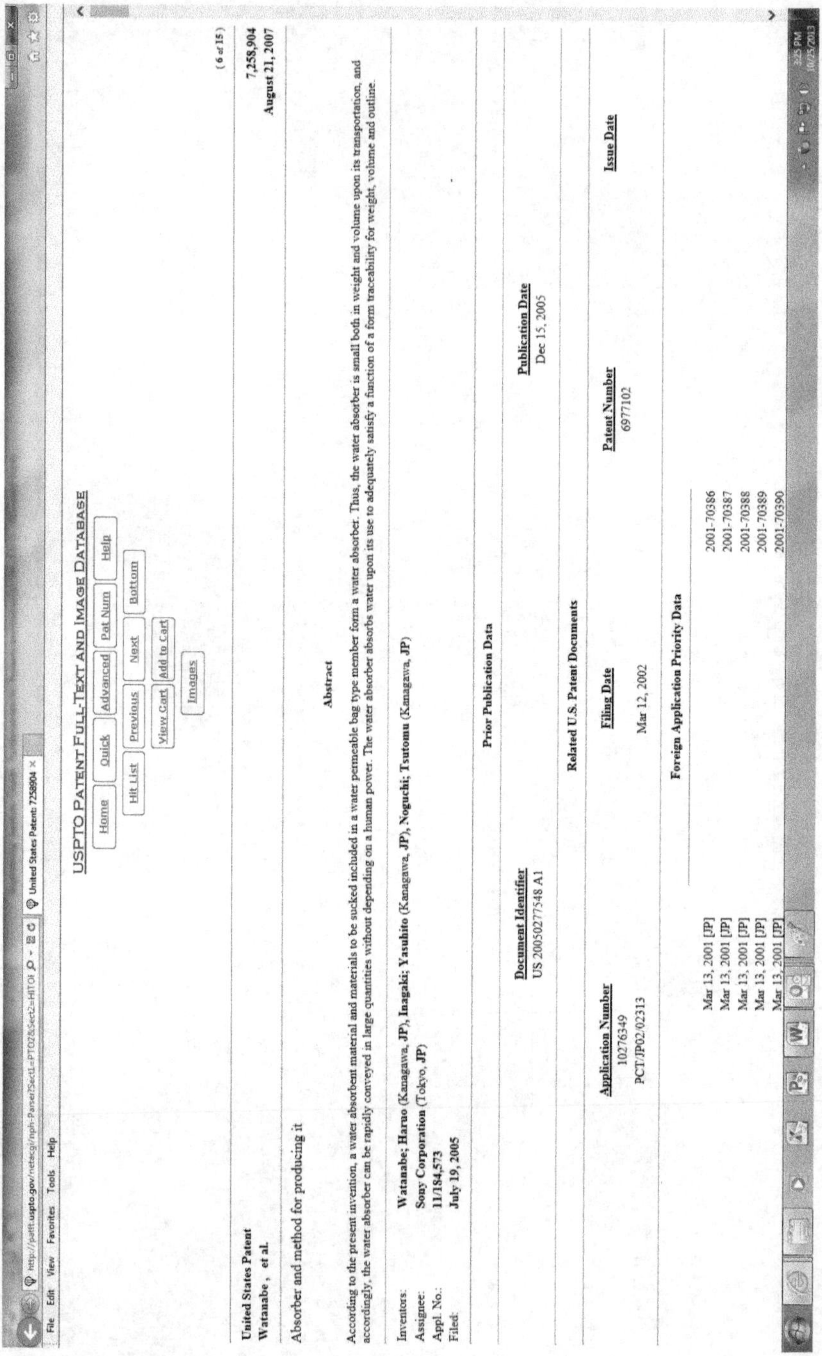

Fig. 7.4 One patent containing the words sandbag, salt, polymer

For a U.S. patent, the USPTO website provides information such as:
- Inventor(s) names,
- Assignee (company or person that owns the patent),
- Related U.S. patent applications,
- Related non-U.S. patent applications,
- Invention claims, and
- A detailed description of the invention that includes drawings and industrial applicability.

The investigator would have to assess if the patent shown in Fig. 7.4 constituted prior art that would prevent the salt/polymer sandbag from being regarded as a novel and nonobvious invention.

Determining patent infringement or whether an invention is patentable is very complex and is best left to a patent attorney. In a capstone class, a team should perform a cursory search of the United States Patent and Trademark Office's patent database to see what similar inventions have been patented. In most cases, the team will not be expected to proceed with an actual patent of the invention. If the team decided to obtain a patent in the United States, the team would engage a patent attorney who would do the following:
- Prepare a summary description of the new invention.
- Describe how the invention would be put into practice, which would include pertinent drawings or schematics.
- Draft one or more claims that describe the invention.

Patenting an invention in the United States only protects it in the United States. To protect the idea internationally a patent application must be applied for in other countries.

Types of Patents

The USPTO website (http://www.uspto.gov/web/offices/ac/ido/oeip/taf/patdesc.htm) issues several different types of patents offering different kinds of protection covering different types of subject matter. The USPTO website defines the following types of patents:
- Utility Patent—Issued for the invention of a new and useful process, machine, manufacture, or composition of matter, or a new and useful improvement of one of the foregoing. The patent permits its owner to exclude others from making, using, or selling the invention for a period of up to twenty years. The vast majority of the patent documents issued by the USPTO in recent years have been utility patents.
- Design Patent—Issued for a new, original, and ornamental design embodied in or applied to an article of manufacture.
- Plant Patent—Issued for a new and distinct, invented or discovered and asexually reproduced plant including cultivated sports, mutants, hybrids, and newly found seedlings, other than a tuber propagated plant or a plant found in an uncultivated state.

Table 7.1 Provisional/Nonprovisional patent application comparison

Provisional patent application	Nonprovisional or utility patent application
An application and not a patent	An application and not a patent
Provides patent-pending status instantly	Provides patent-pending status instantly
Confidential USPTO document *not* available to the public unless (a) the applicant files a nonprovisional application that claims priority to the provisional application, and (b) the nonprovisional application is published or issued as a patent	Confidential USPTO document *not* available to the public unless and until published and/or issued as a patent
Has a non-extendible term of 1 year	The patent has a term of 20 years from its earliest U.S. priority date, and is not renewable
Can be prepared and filed without certain formalities that are required of a nonprovisional application. The filing of a provisional application delays certain expenses associated with the filing and prosecution of a nonprovisional patent application, and the one-year term allows the inventor to further assess the financial and technical merits of the invention	In most cases, it is best to seek the assistance of a patent attorney

- Reissue Patent—Issued to correct an error an already issued utility, design, or plant patent. It does not affect the period of protection offered by the original patent. However, the scope of patent protection can change as a result of the reissue.

The USPTO offers inventors the option of filing a provisional or nonprovisional patent application for an invention. The provisional application provides a lower-cost first application filing and establishes an early effective filing date. It allows the term "Patent Pending" to be applied in connection with a product that embodies the invention. The provisional application has a non-extendible term of one year, and of itself, never becomes a patent. If an applicant who files a provisional application wishes to obtain a patent based on the filing of the provisional application, the applicant must file a nonprovisional application within one year of the filing of the provisional application. Table 7.1 compares features of the provisional and nonprovisional patent application.

Whether the team decides to go forward and patent an invention or not, the research experience and self-education are invaluable to an understanding of the product, process, or service on which the team is working.

Disclaimer The author is not a lawyer. Readers should not act upon any information presented in this chapter without obtaining legal advice from competent, independent, legal counsel.

Epilogue 8

> *We are continually faced by great opportunities brilliantly disguised as insoluble problems.*
> —Lee Iacocca

> *Stay out of the way when the elephants dance.*
> —A corporate axiom

Elephant drawing courtesy of the Wikimedia Foundation. The drawing was released into the public domain by Pearson Scott Foresman.

A Fable

The Blind Men and the Elephant

It was six men of Indostan
To learning much inclined,
Who went to see the Elephant
(Though all of them were blind),
That each by observation
Might satisfy his mind.

The *First* approached the Elephant,
And happening to fall
Against his broad and sturdy side,
At once began to bawl:
"God bless me! but the Elephant
Is very like a wall!"

The *Second*, feeling of the tusk,
Cried, "Ho! what have we here
So very round and smooth and sharp?
To me 'tis mighty clear
This wonder of an Elephant
Is very like a spear!"

The *Third* approached the animal,
And happening to take
The squirming trunk within his hands,
Thus boldly up and spake:
"I see," quoth he, "the Elephant
Is very like a snake."

The *Fourth* reached out his eager hand,
And felt about the knee.
"What most this wondrous beast is like
Is mighty plain," quoth he;
"'Tis clear enough the Elephant
Is very like a tree!"

The *Fifth* who chanced to touch the ear,
Said: "E'en the blindest man
Can tell what this resembles most:
Deny the fact who can,
This marvel of an Elephant
Is very like a fan!"

The *Sixth* no sooner had begun
About the beast to grope,
Than, seizing on the swinging tail
That fell within his scope,
"I see," quoth he, "the Elephant
Is very like a rope!"

And so these men of Indostan
Disputed loud and long,
Each in his own opinion
Exceeding stiff and strong,
Though each was partly in the right,
And all were in the wrong!

So, oft in theologic wars,
The disputants, I ween,
Rail on in utter ignorance
Of what each other mean,
And prate about an Elephant
Not one of them has seen!

The Blind Men and the Elephant, translated by John Godfrey Saxe (1816–1887), is a Hindu fable that occurs in the Udana, a Canonical Hindu Scripture.

A Perspective

The blind men in the fable represent potential team synergy. Individually they do not have a grasp of the situation. If they were to collectively pool their knowledge and abilities, they would have a better understanding of the animal in front of them. Each one of us sees things exclusively from their viewpoint. Each of us could be partly correct but yet, come to a wrong conclusion. We would benefit and gain an improved perspective if we collaborate.

In the same way, the capstone team members bring to the project specific skills—some unique and some overlapping. Collaborative efforts allow a team to
- Tackle more complex problems than they could as individuals.
- Share diverse perspectives.
- Pool knowledge and skills.
- Hold one another mutually accountable.
- Receive support and encouragement from one another.
- Brainstorm and develop new approaches to confront an issue.
- Develop their own voice and perspectives in relation to peers.

The engineering industry frequently uses a matrix structure for development efforts. That is, both functional and product organizations are implemented simultaneously and resources are shared. While there are many benefits, the matrix organization does not exhibit clear lines of authority or responsibility. In addition, a cross-functional team member like an engineer may receive one direction from a functional manager and a different direction from the cross-functional team or project manager. Each part of the matrix operates within the area that they see (or for which they have responsibility). The overlapping reporting relationships can lead to power struggles for resources. Some individuals become disturbed by the ambiguity and conflict that may arise. To avoid conflict, team members stay within their department walls and do not ask broad programmatic questions.

Organizations divide themselves into islands or departments to improve the efficiency of personnel usage. As efficient as it may be, the matrix organization sets up barriers, which limits people's understanding of the whole. The result is much like the fable of the blind men and the elephant - each person's perception is colored by where they stood. And so it is within the organization. Employees working in different departments focus on one specific area of the business while losing site of the big picture (the elephant). Each operational business island may use different operating procedures, different tools, different software applications, and different databases, which are not integrated well with other parts of the organization. The new engineer rarely gets to participate in the entire process, which extends from design conception and working with customers, to developing a specification, to designing, on to product testing, then manufacturing, and training.

Capstone teams participate in a broad experience that integrates business and engineering skills. Team members have the responsibility to make decisions that impact the entire effort. They gain an in-depth understanding of the implication to the bottom line of various decisions they make such as make/buy, purchasing, and vendor selection. Business knowledge and skill are naturally incorporated into project consideration in the students' design, implementation, and test efforts. The results are shown in the form of a written report that summarizes the team's problem analysis, design solution, and ultimate execution of the product, service, or process change. The team makes an oral presentation of its report to an audience of student peers, university faculty, and other stakeholders. They participate in the entire design and development process. They gain a perspective and expertise in activities that few junior engineers experience for several years into their careers.

Appendix: Failure Mode and Effects Analysis (FMEA) and House of Quality

Murphy's Law: If anything can go wrong, it will.

O'Toole's Commentary: Murphy was an optimist.

Use of a Failure Mode and Effects Analysis (FMEA) to Identify a Project Idea

Root Cause Analysis (RCA) is a failure analysis technique that helps multi-disciplinary teams determine the cause of a failure that occurred with a product, service, system or process.

The method uses a Fishbone Diagram, which was first used by Dr. Kaoru Ishikawa. This diagram is used to identify root causes that may have contributed to the problem. An RCA tries to answer the questions:
- What happened?
- Why did it happen?
- What can be done to prevent it from happening again?

RCA helps a team focus on possibilities. Some useful broad failure causal categories that teams consider are:
- Methods or Process
- Materials
- Measurements
- Environment
- Policies
- Equipment
- People
- Working conditions
- Technology
- Management
- Maintenance

The RCA process is a reaction to a failure. The Failure Mode and Effects Analysis (FMEA) is a pro-active method to spot a problem before it occurs. The traditional reason for performing a FMEA is to improve the quality of a process or product quality by anticipating and then preventing a failure. This systematic method helps an organization focus on and understand the impact of potential process or product risks. The FMEA method treats the risk as if the failure had already occurred and corrective action is required.

The FMEA is also an important part of the product and process design activity. The FMEA procedure examines each item in a process, product, service or system, considers how that item can fail and then determines how that failure will affect the overall operation. If the risk of failure may have a detrimental effect on the users of a product, process, service or system the FMEA may offer alternatives for pre-emptive action.

This systematic method assesses the relative impact of different failures, in order to identify the parts of the process that are most in need of change. The FMEA answers the questions:
- What could go wrong?
- Why would the failure happen?
- How serious is the failure?
- What would be the consequences of each failure?
- What corrective actions should be taken?

The approach proposed here uses the FMEA for a different purpose. Engaging in an FMEA exercise may highlight a design idea or a process change that will make significant improvements to an existing system. It requires a team to be familiar with a product, service, process or equipment and its overall operation. The team would then identify potential failure modes and their causes; evaluate the effects of each failure mode on the system; and then identify measures for eliminating or reducing the risks associated with each failure mode. This last step leads to the possibility of developing an idea for a team project. The FMEA is never done in isolation. It requires a multi-disciplinary team's experience and expertise to evaluate potential failures and develop corrective action solutions. The design improvement selected by the team may detect, avoid, or control the problem.

The FMEA process consists of the following steps:
- Analysis of the system with regard to functions, interfaces, and operating modes within the environment in which the unit operates.
- Design and analysis of functional and reliability block diagrams that describe the processes, interconnections, and dependencies.
- Identification of potential failures.
- Evaluation and estimating the likelihood of the failure and its effects.
- Identification of measures to detect and/or prevent errors.
- Evaluation of the effects of suggested corrective action measures.
- Documentation of the results.

The FMEA tool (Tables 1 and 2) assists a team in evaluating each item in a product, service, process or system in a proactive manner. It then asks a team to consider how that item can fail and how that failure will affect the system. The severity,

Appendix: Failure Mode and Effects Analysis (FMEA) and House of Quality

Table 1 Failure mode and effects analysis (FMEA) worksheet

Title:
Date:
Dept.:
Team:

SEV = How severe is effect on the customer, patient, process, product, service or system?
PROB = How frequent is the cause likely to occur?
DET = How probable is detection of cause?
RPN = Risk priority number (RPN) in order to rank concerns; calculated as SEV × PROB × DET

Evaluation of the Potential Failure Mode

No.	Item description	Potential failure mode	Potential effect(s) of failure	SEV	Potential causes	PROB	Current controls	DET	RPN	Recommended action(s)
	Describe the functional operation of the resource.	In what ways can something go wrong?	What is the impact on the customer, patient, product, process, service or system if the failure mode is not prevented or corrected?		What causes the resource to operate incorrectly? (i.e., How could the failure mode occur?)		What are the existing controls that either prevents the failure mode from occurring or being detected it should it occur?		0	What are the actions for reducing the occurrence of the cause or for improving its detection? Provide actions for all high RPNs and on severity ratings of 9 or 10

Prospective problem evaluation

Table 2 Failure mode and effects analysis (FMEA) worksheet

Title:
Date:
Dept.:
Team:

SEV = How severe is effect on the customer or patient or process or system?
PROB = How frequent is the cause likely to occur?
DET = How probable is detection of cause?
RPN = Risk priority number in order to rank concerns; calculated as SEV × PROB × DET

Review and results of resource action

No.	Item description	Potential failure mode	Responsibility and target completion date	Recommended action(s)	Recalculated SEV PRO DET RPN
	Describe the functional operation of the resource.	In what ways can something go wrong?	Who is responsible for the recommended action? By what date should it be completed?	What were the actions implemented? Include completion month/year (then recalculate resulting RPN)	

Results of resource action

Table 3 FMEA severity

Effect	Severity of effect	Ranking
None	No effect	1
Very minor	System operable with slight degradation of performance	2
Minor	System operable with moderate degradation of performance	3
Very low	System operable with significant degradation of performance	4
Low	System inoperable but without impacting process, program, product or service interruption	5
Moderate	System inoperable with slight service interruption without stopping the operation of the process, program, product, service interruption	6
High	System inoperable with moderate process, program, product, service interruption	7
Very high	System inoperable with significant service or product interruption without completely stopping the operation of the process, program, product or service.	8
Extreme	A potential failure mode affects safe system operation, or completely stops the operation of the process, program, product or service execution with warning.	9
Catastrophic	A potential failure mode affects safe system operation, or completely stops the operation of the process, program, product, service execution without warning.	10

probability and detectability of the possible failures represent an estimate at best. Do not seek or expect to get precise values for the proposed failure. Use Tables 3, 4 and 5 as a guide for these values.

Begin completing Table 1 by describing the functional operation of the item or resource. Then list the possible "failure modes" (that is, anything that could go wrong) associated with the described resource. Identify all possible failure causes for each failure mode listed. For each potential failure mode, the team then identifies the effect of the failure on the customer, client, patient, product, process, service or system if the failure mode is not prevented or corrected. For each failure mode, the team assigns a numeric value between 1 and 10 for the for the problem severity (Table 3), for the probability of occurrence (Table 4), and for the likelihood that the overall process or system will detect and report the failure (Table 5).

The Risk priority number (RPN) is the product of the severity, probability and detectability. Assigning RPNs helps the team prioritize areas to focus on and helps in selecting an opportunity for improvement, which might turn into a viable course project.

The technology and health industries use the FMEA tool to identify product, process and system changes that should be implemented. They address high RPN numbers first. In the spirit of continuous quality improvement, the changes actually implemented are evaluated over a three or six month time period with the aid of Table 2. Teams track the RPN number over time to see if changes made to the product, process, service or system lead to a reduced FMEA numbers and quality improvement.

Table 4 FMEA probability of occurrence

Probability of failure	Failure probability	Ranking
Remote: failure is unlikely	No system failure expected	1
	Failures expected in 0.25 % of process, program, product, or service execution instances. Alternatively process, program, product, or service execution downtime is 0.25 %	2
Low: relatively few failures	Failures expected in 0.5 % of process, program, product, service or execution instances. Alternatively process, program, product, or service execution downtime is 0.5 %	3
	Failures expected in 1 % of process, program, product, service or execution instances. Alternatively process, program, product, or service execution downtime is 1 %	4
	Failures expected in 2 % of process, program, product, service or execution instances. Alternatively process, program, product, or service execution downtime is 2 %	5
Moderate: occasional failures	Failures expected in 5 % of process, program, product, service or execution instances. Alternatively process, program, product, or service execution downtime is 5 %	6
	Failures expected in 10% of process, program, product, service or execution instances. Alternatively process, program, product, or service execution downtime is 10 %.	7
High: repeated failures	Failures expected in 20 % of process, program, product, service or execution instances. Alternatively process, program, product, or service execution downtime is 20 %	8
	Failures expected in 30 % of process, program, product, service or execution instances. Alternatively process, program, product, or service execution is inoperable 30 % of the time	9
Very high: failure is almost inevitable	Failures expected in 50 % of process, program, product, or service execution instances. Alternatively process, program, product, or service execution downtime is 50 %	10

Table 5 FMEA detectability

Detection	Detectability	Ranking
Almost certain	Design control will detect potential cause and failure mode	1
Very high	Very high chance that the design will detect the potential cause and failure mode	2
High	High chance that the design will detect the potential cause and failure mode	3
Moderately high	Moderately high chance that the design will detect the potential cause and failure mode	4
Moderate	Moderate chance that the design will detect the potential cause and failure mode	5
Low	Low chance that the design will detect the potential cause and failure mode	6
Very low	Very low chance that the design will detect the potential cause and failure mode	7
Remote	Remote chance that the design will detect the potential cause and failure mode	8
Very remote	Very remote chance that the design will detect the potential cause and failure mode	9
Failure will not be detected	Design cannot detect the potential cause and failure mode	10

Table 6 House of quality

Use of the House of Quality to Identify a Project Idea

The House of Quality is the first matrix in a four-phase Quality Function Deployment (QFD) process. These matrices sequentially relate:
- Customer needs and expectations to technical and performance measures
- Performance measures to features or solutions
- Features or solutions to parts specifications
- Parts specifications to manufacturing processes

It is the most recognized and widely used tool in the QFD method. It is called the House of Quality because of the correlation matrix that takes the shape of a roof sitting on top of the main matrix body—although the roof of the matrix used in this book resembles a right triangle (Table 6) and not an isosceles triangle. This section uses only the first QFD matrix—the House of Quality. Hauser and Clausing [63] discuss the entire process in detail.

Traditionally, QFD is a systematic process used to define customer needs or requirements and translate them into specific plans to produce products and services that meet those needs. It is used by cross-functional teams to identify and resolve issues involved in developing products, processes, and services. The process helps to focus and coordinate an organization's skills to design and then manufacture and market products and services that customers will want.

This section uses the information gleaned from the House of Quality matrix to highlight innovative opportunities for new projects. The process of completing the matrix will promote discussion among the team members and may yield a project idea.

Step 1	Customer Requirements
	Identify and agree on who the customers are. Engage in a marketing effort with the purpose of identifying the customer's real or perceived needs and expectations. What does the customer want?
	The customer's expectations may involve performance, esthetics and usability such as:
	• Color • Appearance
	• Unit or system cost • Shape
	• Operational cost • Ease of use
	• Size • Ease of assembly
	• Weight • Ease of cleaning
	• Speed • Noise
	• Function • Safety
	• Reliability • Accuracy
	• Compatibility with other systems or devices • Capacity
	List as many features as the team wishes. The team may enlarge Table 6 as it contains space for eleven customer requirements. However, a word of warning. Do not get into too much detail because the table can become unmanageably large
Step 2	Customer Importance
	After identifying customer needs, assign a weight ranging from 0 (unimportant) to 9 (absolutely essential) corresponding to each feature suggesting the importance of that feature to the customer and place the number in the *Customer Importance Weights* column. This is a subjective assessment and don't be too concerned about the accuracy of the numbers. Add up the numbers in the *Customer Importance Weights* column and place the sum in the bottom square. The column for the *Weighted Customer Importance* is completed by dividing each assigned importance number by the total and converting to percent
Step 3	Rate the Competition
	As part of the team's foray into marketing, identify competitors. Subjectively compare your product or service versus the competition. Rate both your product/service and the competition's product or service on a scale of 0–9 for each of the customer's requirements identified in step 1.
Step 4	Technical Design Parameters
	The purpose of this section is to translate Customer Requirements into measurable engineering terms. How will the team develop a device or service that meets the customer's needs? The engineering Technical Design Parameters consist of assigning relatively objective metrics, standards or design parameters that meet the customer's need. The metric may include technical characteristics related to performance, size, power, material, weight, strength, surface finish, data rates, interfaces, operating software, failure rate, mean time to repair, manufacturing cost, operating temperature, cooling, etc
Step 5	Relation matrix
	Relations between the customer's requirements and the technical design parameters are not always one-to-one. There are complex relationships and varying levels of strength. A single customer requirement may have an influence on several of the technical design parameters. The relation matrix shows the relationships between "What" (step 1) and "How" (step 4). Examine each Customer Requirement and agree upon a weighting factor which measures the impact of a technical design parameter on the Customer Requirement. Note that the impact may be positive or negative. Assign a weight of 0, 1, 3 or 9 (corresponding respectively to zero, low, medium or strong correlation) to how the customer requirements associate with the engineering parameter.
	An empty column indicates no relationship between the customer and design requirements. The team may find that they need to insert additional technical design parameters to improve the translation from "What" into "How"

(continued)

(continued)

Step 6	Correlation matrix
	The roof or correlation matrix is a triangular shaped table. Team members compare the technical design parameters against one another. The matrix describes the strength of the relationships between the design parameters. The aim is to identify which parameters support each other and which ones do not. Positive correlations mean that the parameter can be increased with minimal or no issues arising. Negative relationships between technical parameters must be examined to eliminate potential problems. The commonly used symbols to indicate the relations between technical parameters are the following ++ Strong positive + Positive − Negative −− Strong negative These symbols are placed in the box at the intersection of the technical design parameters under consideration. Leave the box empty if there is little or no relationship between the technical design parameters.
Step 7	The line indicating the preferred direction represents a simple guide that recommends an increase, decrease, or no change (nominal value) for each of the technical parameters with the intent of providing the best overall product performance. Leave the box empty if there is little or no relationship or the team is uncertain.
Step 8	Perform the following calculations: Raw Score$_n$ = (sum of products in column containing the weighted customer importance x technical design parameter column n) where n = A through L Relative Weight$_n$ = Raw Score$_n$/Sum of Raw Scores, where n = A through L

The resulting relative weight output reflects the amount of customer value residing in each technical component. This gives the team an idea of how much effort they should invest in each component to optimize customer value.

The House of Quality is not simple. The most significant benefit from the process involves promoting discussion among team members. The process "gets people thinking in the right directions and thinking together" [63].

House of Quality Example: Road Bike Drop Handlebar

For bicycle riders, having the right bike handlebar makes a huge difference in the rider's comfort and the bike's performance. The handlebar steers the bike and its shape affects how fast the bike can turn, how far forward the rider has to lean, and where the hands grasp the bars.

The design team is tasked with developing a quality drop handlebar for a road bike. Table 7 shows the resulting House of Quality. Following discussion with many bikers, twelve customer expectations are listed and prioritized. Engineer's meet, discuss and list the technical design parameters (A through L). The team identified four handlebar competitors (Cinelli, Deda, Orange, and Ritchey) and evaluated their comparative strengths. Recognize that this task is subjective. The team agreed upon its competitive goal with price as a main driver. The Relation Matrix correlates the technical design parameters to the Customer Requirements for the team's handlebar implementation.

Table 7 House of quality for a road bike drop handlebar

	Technical Design Parameters											
A	Handlebar diameter is 22.2, 23.8 or 25.4 mm.											
B	Material - carbon fiber, steel, titanium or aluminum alloy	+										
C	Handlebar covering	+										
D	Precision machined											
E	Handlebar width selection 36, 38, 40, 42, 44 or 46 cm	+										
F	Reach selection 75, 80 82 or 84 mm	+	+			+						
G	Handlebar drop selection 130, 135, 144 or 145 mm.					+	+					
H	Tensile properties	+	+			+	+	+				
I	Hardness	+				+	+	+	+			
J	Compressive, Shear, and Torsional Deformation	+	+			+	+	+	+	+		
K	Design Safety Factor							+				
L	Production Cost		++	+	+	+	+	+	+	+	+	
	Preferred Direction (up, down, nominal)	nom	nom	nom	nom	nom	nom	nom	up	up	up	up	down

Customer Requirements & Expectations	Customer Importance Weights (1 to 9)	Weighted Customer Importance (%)	A: Handlebar diameter is 22.2, 23.8 or 25.4 mm.	B: Material - carbon fiber, steel, titanium or aluminum alloy	C: Handlebar covering	D: Precision machined	E: Handlebar width selection 36, 38, 40, 42, 44 or 46 cm	F: Reach selection 75, 80 82 or 84 mm	G: Handlebar drop selection 130, 135, 144 or 145 mm.	H: Tensile properties	I: Hardness	J: Compressive, Shear, and Torsional Deformation	K: Design Safety Factor	L: Production Cost	M	Our Drop Handlebar Goal	Cinelli handlebar	Deda handlebar	Orange handlebar	Ritchey handlebar
1 Comfortable shape	4	5%	9				9	9								7	8	8	8	9
2 Comfortable grip	5	6%	9	3	9		1									6	7	6	7	8
3 Brake and shifter clamps on tight	5	6%	9	9												7	7	8	8	8
4 Snug fit with stem	5	6%	9	9												7	8	8	8	8
5 Comfortable width	8	10%	1				9									8	8	7	5	7
6 Color	8	10%			3	9										7	7	7	7	7
7 Nice finish	8	10%	3		3	9								9		6	9	7	8	9
8 Strong	7	9%	5	9						9	9	9	3	9		5	7	8	9	9
9 Light weight	5	6%	3	9	1									9		7	4	6	7	9
10 Low cost	9	11%	9	9		9							3	9		9	7	6	4	5
11 Comfortable feel	9	11%	3	3	9		9	9	9							8	6	8	8	9
12 Long lasting	8	10%	9							9	9	9	9	3		6	6	8	8	9
Total	81	100%														83	84	87	87	97
Raw Score			4.46	4.25	2.80	1.89	2.40	1.44	1.00	0.78	0.78	0.78	0.59	3.22	0.00					
Relative Weight - %			18.3%	17.5%	11.5%	7.7%	9.8%	5.9%	4.1%	3.2%	3.2%	3.2%	2.4%	13.2%	0.0%					

The outcome of the House of Quality design task represents a guide for designers for a road bike drop handlebar. The results are:

Handlebar diameter is 22.2, 23.8 or 25.4 mm.	18.30 %
Material—carbon fiber, steel, titanium or aluminum alloy	17.50 %
Production Cost	13.20 %
Handlebar covering	11.50 %
Handlebar width selection 36, 38, 40, 42, 44 or 46 cm	9.80 %
Precision machined	7.70 %
Reach selection 75, 80 82 or 84 mm	5.90 %
Handlebar drop selection 130, 135, 144 or 145 mm	4.10 %
Tensile properties	3.20 %
Hardness	3.20 %
Compressive, Shear, and Torsional Deformation	3.20 %
Design Safety Factor	2.40 %

The House of Quality data show that designers should focus on handlebars having multiple diameters, using several different types of materials and a low production cost. Designers then have to balance these requirements with their technical knowledge of the materials involved and the forces applied in this application. Engineers must decide on how the handlebar will deform (elongate, compress, twist) or possibly break as a function of applied load, time, temperature, and other conditions. Remember the House of Quality results are intended to generate ideas. It cannot be taken as the final design word.

References

1. American Society for Engineering Education, The green report—engineering education for a changing world, American Society for Engineering Education (1994) [Online], http://www.asee.org/papers-and-publications/publications/The-Green-Report.pdf. Accessed 6 Jan 2014
2. K.V. Treuren, D. Kirk, T.-m. Tan, S. Santhanam, Developing the aerospace workforce: a boeing experience, in *American Society for Engineering Education Annual Conference & Exposition*, 2010
3. J.H. McMasters, N. Komerath, Boeing-University relations—a review and prospects for the future, in *American Society for Engineering Education Annual Conference & Exposition*, 2005
4. Accreditation Board for Engineering and Technology (ABET), *2014–2015 criteria for accrediting engineering programs* (ABET, Baltimore, MD, 2013)
5. Engineers Canada, *Canadian engineering accreditation board: accreditation criteria and procedures* (Engineers Canada, Ottawa, ON, 2013)
6. J. Sullivan, Talking to teens and the people that influence their lives (2008) [Online], http://wepan.org/associations/5413/files/Tues%20Keynote%20at%20WEPAN%20June%202008.pdf. Accessed 22 May 2013
7. A.S. Thomas, The business policy course: multiple methods for multiple goals. J Manag Educ **22**, 484–497 (1988)
8. D. Moore, F. Berry, Industrial sponsored design projects addressed by student design teams. J Eng Educ **90**(3), 69–73 (2001)
9. R. Pimmel, Cooperative learning instructional activities in a capstone design course. J Eng Educ **90**(3), 413–421 (2001)
10. E. Grubbs, M.W. Ostheimer, Real World capstone design course, in *Proceedings of the 2001 American Society for Engineering Education Annual Conference & Exposition—Session 3232*, 2001
11. D.W. Dinehart, S.P. Gross, A service learning structural engineering capstone course and the assessment of technical and non-technical objectives. Advances in Engineering Education, a Journal of Engineering Education Applications **2**(1) (2010)
12. J. Brustein, Why GE sees big things in Quirky's little inventions (2013) [Online], http://www.businessweek.com/articles/2013-11-13/why-ge-sees-big-things-in-quirkys-little-inventions?campaign_id=yhoo. Accessed 13 Nov 2013
13. J.R. Katzenbach, D.K. Smith, The discipline of teams. Harv. Bus. Rev. **71**, 111–120 (1993)
14. J.R. Dempsey, W.A. Davis, A.S. Crossfield, W.C. Williams, *Program Management in Design and Development* (SAE International, Warrendale, PA, 1964)
15. D. Mackin, TEAMING: tiger teams—the new frontier of teaming (2011) [Online], http://newdirectionsconsulting.com/4608/blog/tiger-teams-the-new-frontier-of-teaming-2/. Accessed 24 June 2013
16. S. Cass, Apollo 13, we have a solution, (2005) [Online], http://spectrum.ieee.org/aerospace/space-flight/apollo-13-we-have-a-solution. Accessed 3 Nov 2013
17. Lockheed Martin Corporation, Skunk works® origin story [Online], http://www.lockheedmartin.com/us/aeronautics/skunkworks/origin.html. Accessed 20 June 2013

18. Lockheed Martin Corporation, Kelly's 14 rules & practices [Online], http://www.lockheed-martin.com/us/aeronautics/skunkworks/14rules.html. Accessed 20 June 2013
19. Fortune Magazine, Six teams that changed the world (2006) [Online], http://money.cnn.com/2006/05/31/magazines/fortune/sixteams_greatteams_fortune_061206/index.htm. Accessed 21 June 2013
20. A. Lashinsky, RAZR'S edge (2006) [Online], http://money.cnn.com/2006/05/31/magazines/fortune/razr_greatteams_fortune/. Accessed 21 June 2013
21. D. Kiley, Ford's new powerstroke: a design breakthrough (2009) [Online], http://www.businessweek.com/magazine/content/09_43/b4152000630377.htm. Accessed 3 Nov 2013
22. S. Musil, William Lowe, the 'father of the IBM PC,' dies at 72 (2013) [Online], http://news.cnet.com/8301-10797_3-57609727-235/william-lowe-the-father-of-the-ibm-pc-dies-at-72/. Accessed 3 Nov 2013
23. B.W. Tuckman, Developmental sequence in small groups. Psychol. Bull. **63**, 384–399 (1965)
24. I.L. Janis, *Victims of Groupthink* (Houghton Mifflin, New York, NY, 1972)
25. B.W. Tuckman, M.A.C. Jensen, Stages of small group development revisited. Group Organ Stud **2**, 419–427 (1977)
26. D.A. Bonebright, 40 years of storming: a historical review of Tuckman's model of small group development, Human Resource Development International **13**(1), 111–120 (2010)
27. J.W. Neuliep, *Intercultural communication: a contextual approach fifth edition* (Sage, Thousand Oaks, CA, 2012)
28. G. Hofstede, *Cultures and Organizations: Software of the Mind* (McGraw-Hill, New York, NY, 1996)
29. G. Hofstede, *Culture's consequences: comparing values, behaviors, institutions, and organizations across nations*, 2nd edn. (Sage, Thousand Oaks, CA, 2001)
30. K. Wilson, Peer and self-evaluation: individual accountability in teams. Landsc Rev **9**(1), 235–239 (2004)
31. B.C. Williams, B.B. He, D.F. Elger, B.E. Schumacher, Peer evaluation as a motivator for improved team performance in bio/Ag engineering design classes. Int J Eng Educ **23**(4), 698–704 (2007)
32. R.A. Layton, M.W. Ohland, Peer ratings revisited: focus on teamwork, not ability, in *Proceedings of the 2001 American Society for Engineering Education Annual Conference & Exposition*, 2001
33. A.F. Osborn, *Applied imagination: principles and procedures of creative problem-solving* (Charles Scribner's Sons, New York, NY, 1953)
34. P. Paulus, V. Brown, Ideational creativity in groups: lessons from research on brainstorming, in *Group Creativity* (Oxford University Press, New York, 2003), pp. 110–136
35. P.A. Mongeau, M.C. Morr, Reconsidering brainstorming. Group Facilitation: A Research and Applications Journal, **1**(1), 14–21 (1999)
36. A.F. Osborn, *Applied Imagination: Principles and Procedures of Creative Problem-Solving—Revised* (Carles Scribner's & Sons, New York, NY, 1957)
37. S. Mannuzza, F.R. Schneier, T.F. Chapman, M.R. Liebowitz, D.F. Klein, A.J. Fyer, Generalized social phobia: reliability and validity. Arch. Gen. Psychiatry **52**, 230–237 (1995)
38. K.K. Dwyer, M.M. Davidson, Is public speaking really more feared than death? Commun. Res. Rep. **29**(2), 99–107 (2012)
39. S. Reeves, Dressing for the job (2005) [Online], http://www.forbes.com/2005/04/28/cx_sr_0428clothes.html. Accessed 26 July 2013
40. J. Oriel, *Guide to Specification Writing for U.S. Government Engineers* (Naval Air Warfare Center Training Systems Division, Orlando, FL, 2013)
41. R. Rai, J. Terpenny, Principles for managing technological product obsolescence. IEEE Trans. Compon. Packag. Tech., **31**(4), 880–889 (2008)
42. Institute of Medicine, *Crossing the quality chasm: a new health system for the 21st century* (National Academy Press, Washington, DC, 2001)
43. Institute of Medicine, *To err is human: building a safer health system* (National Academy Press, Washington, DC, 1999)

44. Institute for Safe Medication Practices, Problems persist with life-threatening tubing misconnections (2004) [Online], http://www.ismp.org/newsletters/acutecare/articles/20040617.asp. Accessed 7 Nov 2013
45. U.S. Food and Drug Administration, Safe Medical Device Connections Save Lives, (2008) [Online], http://www.fda.gov/downloads/medicaldevices/safety/alertsandnotices/ucm134873.pdf. Accessed 7 Nov 2013
46. J. Grout, *Mistake-Proofing the Design of Health Care Processes* (Agency for Healthcare Research and Quality U.S. Department of Health and Human Services, Washington, DC, 2007)
47. A. Reinhardt, Steve jobs: There's sanity returning, (1998) [Online], http://www.businessweek.com/1998/21/b3579165.htm. Accessed 23 Aug 2013
48. R.D. Wesson, Develop an elevator speech. Prism **23**(4), 21 (2013)
49. M. Wats, R.K. Wats, Developing soft skills in students. Int J Learn **15**(12), 1–10 (2009)
50. R.R. Young, Twelve requirements basics for project success, CROSSTALK J. Defense. Software. Eng., 4–8 (2006)
51. J. Gassert, J.D. Enderle, A. Lerner, S. Richerson, P. Katona, Design versus research; ABET requirements for design and why research cannot substitute for design, in *Design versus Research; ABET Requirements for Design and Why Research Can Proceedings of the ASEE Annual Conference and Exposition*, ed by J. Gassert, J.D. Enderle, A. Lerner, S. Richerson, P. Katona (Chicago, 2006)
52. M.E. Conway, How do committees invent?, *Datamation* **15**(5), 28–31 (1968) [Online], http://www.melconway.com/research/committees.html. Accessed 20 June 2014
53. J. Frederick, P. Brooks, *The Mythical Man-Month: Essays on Software Engineering* (Addison Wesley Longman, Inc., New York, NY, 1995). Anniversary Edition
54. J.D. Gassert, J.D. Enderle, *Design versus research in BME accreditation*. IEEE Eng Med Biol Mag. **27**(2), 80–85 (2008)
55. J.M. Stecklein, J. Dabney, B. Dick, B. Haskins, R. Lovell, G. Moroney, Error cost escalation through the project life cycle, *National Aeronautics and Space Administration* (Johnson Space Center, 2004)
56. S. Canright, NASA engineering design challenge (2008) [Online], http://www.nasa.gov/audience/foreducators/plantgrowth/reference/Eng_Design_5-12.html. Accessed 2 July 2013
57. National Aeronautics and Space Administration (NASA), NASA Engineering Design Challenge [Online]. http://www.nasa.gov/audience/foreducators/plantgrowth/reference/Eng_Design_5-12.html#backtoTop. Accessed 8 July 2013
58. J. Russell, statics and engineering design—a new freshman engineering course at the U. S. Coast Guard Academy, in *32nd ASEE/IEEE Frontiers in Education Conference*. (Boston, MA, 2002)
59. M.C. Paretti, J.J. Pembridge, S.C. Brozina, B.D. Lutz, J.N. Phanthanousy, Mentoring team conflicts in capstone design: problems and solutions, in *American Society for Engineering Education Annual Conference* (Atlanta, 2013)
60. R.E. Larson, D.A. Miller, A case study in capstone organization for continuous design/build projects, in *American Society for Engineering Education Annual Conference* (San Antonio, 2012)
61. J.P. Kotter, *A Sense of Urgency* (Harvard Business Press, Boston, MA, 2008)
62. J. Welch, S. Welch, The 6 deadly sins of leadership (2013) [Online], http://www.linkedin.com/today/post/article/20130327154206-86541065-the-six-deadly-sins-of-leadership. Accessed 22 Aug 2013
63. J.R. Hauser, D. Clausing, House of quality. Harv. Bus. Rev. 3, 63–73 (1988)

Index

A
Accreditation Board for Engineering and Technology (ABET), 5, 88
Agenda, 26, 27, 31, 33

B
Brainstorming, 3, 28–30, 45, 89, 90, 95, 96, 125
Breakthrough, 43
Budget, 1, 4, 8–10, 19, 39, 50–58, 60–63, 65–67, 77, 80, 81, 83, 96, 97, 100, 104, 110, 111
Budget template, 52

C
Canadian Engineering Accreditation Board, xiv, 139
CATME, 22
Communications, 3, 4, 16, 21–23, 25–38, 47, 69, 70, 73, 88, 99, 104, 105, 111
Conducting meetings, 26
Confucius, 103
Copycat, 42
Copyright, 46, 50, 114
Critical Design Review (CDR), 65–66

D
Design process, 5, 87–90
DRIDS-V, 95

E
Engineers Canada, xiv, 139
Engineers Without Borders, 2
Ethics, 2, 21–22
Evaluation, 14, 22–24, 29, 83, 85, 89, 92, 95, 98, 128, 129

F
Failure Mode and Effects Analysis (FMEA), 44, 80, 85, 102, 127–136
Final report, 3, 40, 51, 72, 74, 76, 80, 83–85
FMEA. *See* Failure Mode and Effects Analysis (FMEA)
Formative, 24

G
Groupthink, 14, 15, 30

H
Hofstede, Geert, 19–21
House of Quality, 85, 102, 127–136

I
Ilities, 100, 101
Intellectual property, 46, 50, 51, 59, 73, 74, 81, 89, 102, 103, 109, 113–122

J
Johnson, Kelly, 10, 12

K
Key performance indicator (KPI), 52, 66, 76, 77

L
Leapfrog, 42
Luer connectors/fittings, 44

M
Marketing, 2, 8, 9, 48, 51–53, 57, 59, 71, 78, 99, 100, 103, 133, 134
Microsoft Project, 4, 58, 73
Minutes, 26–28, 33, 54, 55

N
National Aeronautics and Space Administration (NASA), 9, 88–91, 95

P
Patent, 3, 41, 46, 50, 54, 74, 81, 82, 96, 100, 109, 114–122
Peer evaluation, 22, 24
Piggyback, 42
PowerPoint, 31–38, 40, 47
Preliminary Design Review (PDR), 65–66
Primavera, 4
Product life cycle, 73, 101
Project selection, 40, 42–48, 109
Proposal preparation, 12, 40, 48–86
Proposal report, 66, 67, 71–83

Q
Quirky, 5

R
Risk, 4, 9, 13, 20, 53, 55, 56, 58, 60–66, 72, 76, 79, 82, 83, 85, 86, 91, 96, 101, 102, 110, 111, 115, 128–130, 132
Root Cause Analysis (RCA), 127, 128

S
Six Sigma, 44
Skunk works, 9–12
Specification, 1, 3, 4, 8, 10, 11, 13, 23, 50, 51, 55, 57, 59, 63, 65–71, 74–76, 80, 82–86, 88–95, 97, 98, 100, 102, 110, 125, 132
Stakeholder, 5, 21, 45, 54, 58, 63, 66, 73, 76, 82, 85, 86, 111, 112, 125
Statement of Work (SOW), 71–73
Strength, Weakness, Opportunity, Threat (SWOT), 45–48, 51, 59
Summative, 24

T
Test, 2–4, 9, 10, 13, 17, 23, 45, 49, 51, 53, 59, 61, 62, 65–67, 74–76, 81, 84, 86, 89, 91–95, 97, 98, 100, 101, 104, 109, 110, 125
Tiger team, 9–10
Trademark, 10, 46, 50, 114, 116, 117, 121
Traditional team, 8–9
Tuckman, Bruce, 13, 15, 30, 47

V
Validation, 69, 90, 91
Verification, 17, 51, 69, 74, 76, 81, 84, 86, 90–95, 98, 104

W
Weekly status report, 12, 56, 58–61, 65
5 Whys, 17, 18
Work breakdown structure (WBS), 23, 50, 51, 55, 58, 59, 76, 97